INHERITANCE OF CREATIVE INTELLIGENCE

Jon L. Karlsson, Ph.D., M.D.

INHERITANCE OF CREATIVE INTELLIGENCE

Nelson-Hall, Chicago

Library of Congress Cataloging in Publication Data

Karlsson, Jon L.
   Inheritance of creative intelligence.
   Bibliography: p.
   Includes index.
1. Intellect—Genetic aspects.
2. Creative ability—Genetic aspects.
3. Human genetics.
4. Mental illness—Genetic aspects.
I. Title.
BF431.K385   155.7   77-19297
ISBN 0-88229-391-5 (cloth)
ISBN 0-88229-607-8 (paper)

Copyright© 1978 by Jon L. Karlsson

All rights reserved. No part of this book may be reproduced in any form without permission in writing from the publisher, except by a reviewer who wishes to quote brief passages in connection with a review written for broadcast or for inclusion in a magazine or newspaper. For information address Nelson-Hall, Inc., 325 West Jackson Blvd., Chicago, Illinois 60606.

Manufactured in the United States of America

10   9   8   7   6   5   4   3   2   1

*This book is dedicated to
the many creative scholars who remain
unrecognized despite their superior
judgment and wisdom.*

Table of Contents

Preface   ix

Section I   HUMAN GENETICS AND ITS METHODS
1. Genetic Diversity in Human Populations   1
2. Pedigree Studies of Common Genetic Systems   9
3. Quantitative Family Studies   17
4. Estimates of Heritability   25
5. Identification of Useful Mutant Genes   33
6. Biochemistry of Gene Function   39

Section II   GENETIC TRANSMISSION OF GIFTEDNESS
7. Definitions of Intelligence   47
8. Hereditary Contributions to Learning Ability   53
9. Evidence for a Myopia Gene   63
10. Myopia and Intelligence   73
11. Existence of an Alcoholism Gene   79
12. Brain Stimulation Associated with Alcoholism   87
13. Physiology of Learning Functions   93

Section III   BIOLOGIC FACTORS IN CREATIVITY
14. Definition of Creativity   97
15. Evidence for Inheritance of Creativity   103
16. Existence of a Schizophrenia Gene   107

17. Genetic Transmission of Psychotic Tendency   121
18. Relation of Creativity to the Schizophrenia Gene   131
19. Physiologic Regulation of Brain Arousal   143

**Section IV PERSONALITY GENETICS AND
SOCIAL PLANNING**

20. Inherited Personality and Creative Aptitude   155
21. Continued Integrity of Man's Genetic Heritage   167
22. Future of Creative Intelligence   179

References   191

Index   201

Preface

Since man occupies his unique position in the world mainly as a consequence of his superior brain, the subject of its operation is naturally of great interest. There has been much debate about the exact importance of genetic factors in human intelligence, but people have always felt that inherited differences play a role, some persons being born with better mental endowment than others. Recent concern about declining scores on scholastic aptitude tests, administered to prospective college students, also make timely a reconsideration of the evidence for heredity in human intellectual development.

Human genetics, in actual practice, has for a long time dealt primarily with abnormalities or diseases, the tools for an approach to a study of normal or superior development being, therefore, relatively undeveloped. The basic concepts are, however, the same, dependent on the transmission of individual pairs of genes, each chromosomal locus being responsible for a particular biochemical process. While abnormalities are usually easy to recognize, the normal counterparts are frequently less visible. Diseases can thus often be identified with particular genes, whereas studies of normal characteristics and their range of variation remain less specific in nature.

The present volume is the culmination of years of research in human genetics, many of the concepts having evolved gradually as the underlying factors unfolded with the accumulation of additional knowledge and acquisition of new insight. To acquire the feel for a field one must live with it for many years, and usually it is difficult later in life to master the necessary background and truly develop a working facility with new methods. Since I have devoted my entire adult life to a study of genetics, this exposure has given me certain advantages in the sense of being "conversant" with the field. Originally I studied basic genetics at the University of California at Berkeley, emphasizing the area of mammalian inheritance. Later I did graduate research in microbial and biochemical genetics, before taking out time to acquire a medical degree and specialize in pediatrics. My present interests are mainly in the area of growth and development, always considered relevant to pediatrics. I have applied my education to the study of human abnormalities, more recently turning to the inheritance of personality differences. Obviously there are few scientists with a background parallel to mine, and therein lies the justification for my attempt to make the present contribution.

A task of this nature must usually be undertaken, in relative isolation and at some sacrifice, by a devoted scientist. Social factors generally operate in the direction of maintaining the status quo, probably because current leaders do not see it in their interest to support a change or get involved in controversial issues. While I was still young and naive, I assumed that encouragement would soon come if one proved oneself productive in an obviously important area, but I learned through experience that this is not the way the real world operates. If I had not been in a position to support my effort by doing medical work, both in terms of necessary financing and access to the clinical material, none of this research would have come to fruition.

An attempt to tackle the problem of inheritance of intelligence was first conceived by my brother, Dr. Askell Löve, who at that time was Professor and Chairman of the Department of Biology at the University of Colorado. He was, of course, aware of my special qualifications, and he unselfishly continued his support and encouragement when questions and adversities arose. Discussions with my former teacher, Professor Emeritus Everett Dempster of the University of California, have also been useful and reassuring. Mrs. Alberta Titus and Miss Catherine Baker, librarians of the Napa State

Hospital, assisted me in locating the necessary literature. My wife, Lura Karlsson, has done the typing and editing as well as assisting with the data collection. I am also thankful to Mr. Tom Fotinos, Principal of the Napa High School, for making student records available and to Professor Tomas Helgason, University of Iceland, for access to the files of mental patients. Additional help with genealogic information was given by Mr. Bjarni Vilhjalmsson, Director of the National Archives of Iceland.

Mr. William Steubing, Acquisitions Editor with Nelson-Hall, made the suggestion that I write this book, and he and his staff have given the effort every support.

J. L. K.

SECTION I
Human Genetics and Its Methods

1
Genetic Diversity in Human Populations

After a long period of stagnancy and vacillation in the study of inheritance of human personality traits, an air of expectancy and optimism is finally emerging in this important field. At last there is some hope that a breakthrough may be around the corner, capable of reversing the hesitancy in regard to this subject which has for too long characterized the attitude of human geneticists.

In reality there was never a reason for the timidity which almost paralyzed the leaders in behavioral genetics, causing them to back off apologetically in the face of aggressive criticism from proponents of environmental theories. Although the latter were quite willing to loudly proclaim supremacy for their boldly tailored postulations, explaining through external influences the development of human behavior and intelligence, their arguments actually amounted to little more than speculations or personal opinions.

Importance of heredity in human development

The evidence for hereditary influences in mental development has always been compelling, even if it had some inadequacies and needed to be bolstered further. In part the weaknesses of the available data were engineered by the critics, since they controlled the funds needed for further research by steadfastly refusing to let money be used for the collection of information that might weaken their grip on the established empire.

It has not been uncommon for investigators in genetics to be denied financial support, for example on the grounds that a proposed investigation was too difficult to carry out and therefore not worth attempting. Then the same officials could justify their continued castigation of the genetic evidence by claiming that the data were inadequate. Unfortunately it is in the nature of governmentally sponsored institutions to perpetuate their own existence. If they are founded for the purpose of pursuing new knowledge, their two goals may at times be in conflict, survival usually winning out. With their leaders incapable of resisting the pressures, the few geneticists interested and capable in the area of behavioral development were left defenseless.

The present air of optimism in personality genetics does not stem from a change in the official attitude or from emergence of unexpected support; rather it has its roots in new interpretations of a basic nature. There has been a reluctance to accept the very fundamental position of the evolutionary concept in the determination of human characteristics, but now suddenly there is a realization that the facts have been there for some time and the conclusions are waiting to be drawn. Consequently, it turns out to be mainly a change in mental attitude that was needed to overcome the artificial obstacles and allow the real picture to be seen in proper perspective. Hopefully these developments will lead to an official recognition of the need for increased emphasis in this area of endeavor.

To effectively pursue investigations founded on evolutionary concepts, it is necessary to approach the problem with an open mind and a willingness to consider possibilities which may diverge from ideas considered valid by scholarly leaders of the past. The evolutionary theory of man's origin is still debated in some circles, but few dispute the thesis that genetic changes and biologic adaptation indeed take place, and this is all that needs to be assumed to follow the arguments presented in this book. If man has progressed from a relatively unsheltered food-gathering existence to his present dependence on a highly organized industrial society, it seems inevitable that biologic changes have accompanied the adjustments and that differential survival has led to a selection of individuals best suited to a complex modern-day life. Such adaptation carries with it changes in biochemical balances, a different combination of hereditary units or genes being favored when the environment is altered, perhaps at times in response to new demands that man himself places on his surroundings.

Some will feel that the worth of the individual is being questioned or even degraded by a serious consideration of a biologic superiority of certain members of the human community. But differences are a fact of life which will not disappear, even if some professionals see fit to ignore them. The exploration undertaken in this book demands a willingness to face the observable realities, although they at times may turn out to be contrary to the wishes of well-meaning scholars. Denial of the facts only perpetuates the stale state of affairs which already has dominated this area for too long, encouraging complacency when positive action may be needed to protect and preserve man's biologic heritage.

Physical basis of human evolutionary progress

The observation has been made by students of anthropology that the ancestors of modern man appear to have displayed a burst of evolutionary progression once they embarked on a new life on grassy plains, adequately supplied with large animals which formed an abundant food source. This new existence demanded cooperative hunting, rather than individual food gathering, and in turn a premium had to be placed on creativity in developing new approaches to life, including skills in verbal communication. Such factors are presumed to have served as catalysts for increased use of the brain, in the design of hunting procedures, in the making of weapons and tools, and eventually in the building of permanent shelters. The fossil evidence indicates that his new mastery of the environment then enabled man to move between continents, adapting himself to different circumstances, and gradually occupying the entire world. This process culminated in the development of agriculture and industry, with a parallel evolution of human culture.

Special explanations are demanded by the puzzle of the extraordinarily rapid genetic changes exhibited by early man. In a span of just two million years he was somehow able to emerge from a near-animal existence to a complex level of industrialization and culture. The usual genetic mechanisms do not manifest sufficient flexibility to permit such speedy development.

The principles governing human biologic progress have actually been fairly well defined, being in essence equivalent to mechanisms of genetic adaptation in other organisms. Although the course of organic evolution is founded on gene mutations, it is best visualized by thinking of regulation of chemical rates. While there is still some debate about the exact source of the building materials utilized in the evolutionary process, it now seems clear that a major reservoir of the basic units is genes which have undergone a small alteration in structure, thus being programmed to make possible a slight upward or downward adjustment in the rates of chemical reactions. A mouse and a man possess all the same organs and essentially identical metabolic equipment, but presumably reaction rates differ, leading to different end products during the process of growth and development. Evolution to the higher form has not been based on radical changes, but rather on a series of small adjustments, eventually followed by a sterility barrier to establish a new species no longer capable of interbreeding with others.

Long-range genetic adjustments are thus dependent mainly on a gradual accumulation of small changes in the hereditary material, rather than relying on sudden creation of new genes or on gross alterations in the function of preexisting factors. They cannot be based primarily on the more commonly identified disease-related nonfunctional kind of mutant gene, which is thought to lead to a loss of a step in a biochemical reaction sequence, although at one time there was confusion on that point. This does not mean, however, that the nonfunctional type of gene cannot serve some evolutionary function, and it appears to be the most readily available altered form of the genetic material. Quick adaptive changes, which cannot await the accumulation of slowly forming, precisely tailored genes, can probably at times be accomplished through adjustments of the balances between older units, incorporating the metabolically inactive type of gene when that proves advantageous to the organism.

Group adaptation in human evolution

The origin and persistence of a higher level of variation in man, especially in more developed cultures, is thought to have its roots in human group living and communication ability. The possibility of transmitting information and thoughts to other members of the species has created an opportunity for specialization and division of labor, individuals with unusual abilities being an asset to the human community, even if they may themselves at times be burdened with compensatory detrimental traits. In species where each individual operates to a large degree independently, all must possess the attributes essential for survival. However, in a hunting group, or later in a culturally advanced human society, a highly specialized member may in fact have found an improved opportunity and have been protected by his peers, even if he lacked some of the characteristics necessary for independent existence. For example, Albert Einstein was a peculiar man, seemingly unmotivated to care for some of his basic needs, yet his special brain powers made him probably the most valuable single member of his society. Benefits derived from the brain power of one individual may not be limited to a small group, but can extend to an entire nation or race in terms of survival advantages. Indeed, the gains may be utilized by future generations through cultural heritage.

The adaptation to a cooperative hunting existence and the call for special talents and ingenuity on the part of modern society, which can be viewed as reflections of group adjustment as opposed to individual adaptation, have created the possibility of utilizing expedient genetic mechanisms, which to some extent are denied organisms lacking effective communication skills. As was stated above, the long-range evolutionary processes are probably dependent largely on relatively small alterations in the genetic material, which adjust biochemical reaction rates to levels that are optimal under new circumstances. In the overall biochemical machinery, only some of the chemical steps are rate-limiting, and most of the genes probably cannot be improved upon; they are already doing exactly what they are needed for. Besides, only certain reactions are adjustable so as to produce improvement; under discussion here are those processes which in some manner lead to enhancement of brain power.

Assuming that some chemical steps indeed exist, where a change in the reaction rate, either toward a slowdown or a speedup, could be beneficial, the ordinary evolutionary processes would be extremely sluggish in responding to such a need. Furthermore, if the requirement for a change can be met by a few individuals acquiring special powers which they are capable of communicating to the group, it may not be advantageous to the species as a whole to make an adjustment to a different chemical rate, which presumably would have to occur through a small change in a particular gene. Specific gene changes also require thousands of years for the proper gene form to show up by chance and become established. In addition, if the adjustment called for is in response to a current need, which may no longer exist at a later time, the organism might enter a blind evolutionary alley by making a total adjustment to the situation at hand.

Hybrid vigor and related systems

This is where the opportunity arises for the use of alternate genetic pathways, some of which can be seen as temporary in nature, not calling for a total shift of the entire gene pool. Such mechanisms, including so-called hybrid vigor, often involve a price, some members having to be sacrificed in each generation for the overall good of the species. A chemical rate can be adjusted, probably mainly downward, by the utilization of a nonfunctional type of gene, which always exists in the gene pool. Hence the heterozygous carrier, with only one functional gene of a given type, may be just what is called for to accomplish the necessary task for the time being. If the carrier exhibits a more optimal chemical rate than the noncarrier, the former would be favored by the forces of natural selection, increasing the frequency of the nonfunctional gene and establishing a new balance. There could also exist some chemical processes whose total abolishment, previously blocked by associated abnormalities, has become feasible despite its former price, because of new developments, such as a change in climate or invention of a medical device. Here, too, a nonfunctional gene would show an increased frequency, being in reality in the process of replacing the original gene. Such shifts are, of course, dependent on a competitive existence, in which only the most fit survive to reproduce. Although in these situations a chemical rate is presumably adjusted downward, the net result may actually be an increase in a specific metabolite.

One may now wonder how this type of genetic shift, utilizing nonfunctional mutant genes, would manifest itself in the overall economy of the species. The better-adapted heterozygote may not be so radically altered as to stand out from the other members. But the increased frequency of the nonfunctional gene may result in readily apparent illness in a significant number of heterozygous carriers.

Genetic Diversity in Human Populations

Abnormal homozygotes, having two nonfunctional genes, are also multiplied in number, many showing a severe disease. The apparent result is thus an increase in members with an inborn disorder or disease tendency, and this is likely to be what catches the eye.

Looking now at the proposed system from the vantage point of a geneticist trying to explain evolutionary processes, one may ask whether those genetic diseases which are known to occur at high frequencies in present day human populations have not in reality built up to their current levels via mechanisms of the type described above, such as those involving hybrid vigor. The obvious way to attempt to answer that question is to search for beneficial concomitants associated with those genetic disorders which persist at high levels. It indeed seems probable that some disorders have multiplied in frequency by way of such systems. These disorders could then be viewed as the price man has had to pay for his rapid biologic adaptation in recent times.

Natural selection mechanisms

Besides the operation of the proposed systems utilizing defective types of genes, it seems probable that some of the observed variation in human mentality may have been brought about by genes which are only slightly altered from the original form, rather than being grossly deficient. One can, however, reason that an entirely favorable gene, perhaps showing some beneficial effect in the heterozygous state and even greater effect in a homozygous condition, would already have established itself in most of the population, only such genes remaining permanently at a lower rate as have a price connected with their multiplication. Since outstandingly brilliant individuals are rare, they may well arise when infrequent combinations occur of genes which for some specific reason are held in check, so that they continue to exist at relatively low rates.

The working hypothesis elaborated above may be restated more succinctly as follows: One can first visualize a basic human type, exhibiting relatively little variation and adapted to nonspecialized kinds of occupations. This would require each individual to be well fitted for an agricultural or hunting existence, without great need of intellectual powers. In societies where inventiveness and specialization have become necessary, the basic type would be increasingly supplemented by individuals with enhanced abilities in limited areas, often attained at a sacrifice by such persons of the total adaptation to a more primitive kind of life style. Although limited in certain respects, these persons would confer increased fitness on their societies. One can imagine that highly specialized individuals, having multiplied in number only in relatively recent times, are dependent on genes which often have detrimental effects in addition to the evident brain stimulation. Hence the overall population becomes burdened with a compensatory increase in genetic diseases of high frequency.

It is thus the principal thesis of this book that important aspects of the rapid recent progress in human evolution have been dependent on mechanisms which produce for temporary purposes changes associated with a sacrifice in another area. This shows up as a heavier burden of certain diseases, and in this manner the genetic shift becomes apparent. It is conceivable that this mechanism may in reality account for a large portion of the great intellectual diversity existing today in human populations. Group evolution, based to a large extent on the development of culture, is then dependent on the occasional emergence of a few individuals of superior mentality, the progress of their societies in essence hinging on their contributions. These concepts also have far-reaching implications in terms of natural balances that mankind must have adapted itself to, and these cannot be disturbed without serious consequences.

Proceeding on this basis, one may search for explanations of important differences in human mental abilities. This exploration is in an area of endeavor which is likely to assume increasing importance as man continues to interfere with mechanisms of natural selection. Differential survival and reproduction probably have in the past not only insured continued progress, but also guarded against a loss of the biologic gains already made.

2
Pedigree Studies of Common Genetic Systems

    Mendelian genetics as a scientific discipline is only a century old, so that many of its principles are still in a stage of development. A link of genetics with biochemistry is even more recent, the chemical structure of the genetic material having been under intensive investigation for just the last quarter century. The exact physiologic operation of the genetic machinery, in relation to fine adjustment of regulatory mechanisms and minor variability of chemical reaction rates, is just beginning to be explored.

    The basic concepts underlying these processes are well defined, and some familiarity with them is necessary, before it is possible to meaningfully discuss the genetic basis of the normal ranges of intellectual variation.

The nature of man's basic hereditary material

    The thousands of different genes carried on the chromosomes in the nucleus of each cell are recognized to be the determinants of all hereditary characteristics. Except for males having just one member of all sex-linked genes located on the X chromosome, normal persons generally possess two allelic partners at each chromosomal locus, one having been contributed by either parent.

    By controlling the production of a polypeptide, a constituent of an appropriate enzyme, a genetic locus usually governs a specific chemical reaction in the body machinery. Enzymes are biochemical catalysts which regulate all chemical processes in the body. Even when the two genes at a particular locus are both metabolically active, they may nevertheless differ from one another, the chemical reaction rates mediated by the separate "codominant" partners perhaps showing small differences. The presence of a slightly altered form of a gene, whether in single or double dose, seldom leads to a disease condition, but codominant mutant genes, presumably by making possible a more optimal adjustment of the corresponding reaction rates, are thought to be the chief source of raw material for the long range evolutionary process. Although they are thus recognized as very important, codominant genes are in practice difficult to study, because of their slight differences from the ancestral or native genes.

All genes periodically undergo random or spontaneous mutations, this being the process which ensures a constant supply of genetic variability. The most likely changes appear to be those which cause total interference with the function of the gene. All metabolically nonfunctional forms of a gene, although not necessarily alike, are for genetic purposes identical. When they are lumped together, they often constitute the classical "recessive" gene, which then forms the counterpart of the "dominant" functional form, applying these terms as they are ordinarily used by geneticists. While a more detailed study, employing refined techniques, can sometimes identify different subvariants of the opposing dominant or recessive types of gene, it is sufficient for the purposes of the present discussion to think of genes as existing mainly in two major forms, one active and the other inactive, the latter often being associated with a disease state. In reference to the recessive factor, the expression "gene frequency" denotes the fraction or percentage of all the genes at a particular locus which exists in the altered form.

In regard to their overt manifestation, disease-producing genes indeed appear to be identical in different families. Presumably in the majority of instances they are of the metabolically nonfunctional variety, having arisen through mutations which cause sufficient alteration to the structure of the gene to render it incapable of guiding the production of the appropriate enzyme. In genetic studies of experimental organisms, such genes generally are likely to be classified as recessive, as was stated above. However, the terminology in human genetics has evolved differently, permitting many metabolically nonfunctional genes to be considered dominant. The chief reason for this difference is the unavailability of breeding experiments in human genetics, leaving rather inaccessible the information on the homozygote with two abnormal genes. Besides, minor deviations or disease conditions in heterozygous carriers of one mutant gene are more likely to be noted in man. Thus the mutant form of a gene may result in a recognizable medical condition in the heterozygote, without knowledge being available about the rare homozygote. Such a gene in human genetics then fits the special definition of dominance applied to man, the disorder being regularly passed from an affected parent to one-half the offspring, although it may not meet the more rigorous criteria for dominance in other species. Assuming that the homozygote, if known, would indeed show a more severe disease, a corresponding gene might in animal genetics be considered recessive, if the mild disorder of the heterozygote remains undetected. At most it would be classified as intermediate in expression, when the heterozygous carrier differs from either homozygote. The group of dominant disorders in man thus embraces many genes that in other species would be placed in the intermediate or recessive categories.

If a metabolically nonfunctional gene causes no abnormality in the heterozygous carrier, the gene can only be classified as recessive, leading to a disease exclusively in the abnormal homozygote. The disease, however, is likely to be severe, as the corresponding chemical reaction is totally arrested. In this situation the native or functional gene constitutes the dominant counterpart, persons who possess either one or two such genes being equally healthy.

Simple genetic mechanisms

The principles of genetics were originally developed around "unit characters," such as flower color in plants, where the occurrence of opposing dominant and recessive traits can be readily demonstrated. In human genetics it is also possible to select corresponding situations, for example the inheritance of brown versus blue eye color in Scandinavian populations. Persons are brown eyed if they either inherit two genes for brown, *BB,* with a functional gene from each parent, or if they possess just one active gene, *Bb*, with the native gene matched up with a mutant or nonfunctional one. The recessive genotype, *bb*, forms no brown pigment, thus manifesting the basic blue color that all normal persons possess.

Now that a great deal more is known about genetics, it is clear that most genes cannot be described by use of opposing readily apparent characters. The native gene, through the production of an enzyme, regulates a biochemical process, which usually comprises just one step in a chemical sequence, the latter in turn being interlaced with other chemical phenomena. There may thus be no identifiable dominant trait corresponding to the disorder which appears when the chemical reaction is arrested. Still it is useful to think in terms of simple hereditary mechanisms as a starting point in genetic research, maintaining an awareness that this is likely to be just the first step toward a more comprehensive analysis of the problem.

Until recently, no quantitative procedures have existed which were specifically designed for estimating the importance of hereditary factors in the development of an observable human characteristic, the operation of a genetic system often being deduced instead from an inspection of family pedigrees. Some conditions are seen to be distributed in families according to the Mendelian laws and thus are evidently of genetic origin. This is true of any trait regularly passed from an affected parent to 50 percent of the offspring, illustrating the classical dominant type of transmission. A disorder occurring in 25 percent of the sibs of index cases, when the parents are unaffected, establishes a recessive mechanism. The third regular system, sex-linked inheritance, involving a gene located on an X chromosome, leads to passage through unaffected mothers to one-half of the male offspring, and here, too, no one is likely to dispute the existence of a genetic etiology.

Although these proportional relationships seem very definitive, there are often problems in human genetics when it comes to actually establishing the occurrence of precise numerical ratios. If sufficient pedigree material is available, perhaps covering several generations, straightforward dominant or sex-linked inheritance should not be hard to recognize. However, dominant transmission can occasionally be mimicked by a frequently occurring recessive disorder, leading in some families to direct passage through several generations and affecting one-half of the children in certain segments. This has been referred to as pseudodominance and has sometimes resulted in a disorder being placed in the wrong genetic category.

Recessively inherited disorders can also be difficult to properly document, and even more so when the frequency of the abnormal gene is high. Vulnerable families come to attention only when an affected member appears, and with a one-in-four risk, some such families are by chance skipped. Once an index case appears in the offspring of carrier parents, the risk in the subsequent children is one-fourth, but the data are often quite scanty if one eliminates from consideration the index cases and all previously born unaffected members. Methods are available, however, which permit use of all the material in affected families, correcting for the distortions described above. The ultimate test for pure recessive transmission is to demonstrate that matings of two affected individuals invariably lead to nothing but affected offspring, as no other mechanism produces this result. However, many recessive disorders are associated with a severe disease condition, limiting the reproductive capacity and also maintaining the disorder at a low frequency, mating of two affected persons then being quite rare.

Investigators must be aware that the observed relationships are not always simple, even if the basic mechanism is a straightforward one. Selection bias can also alter the findings, for example by leading to overrepresentation of multiple case families when the research material is located through hospital records. Appropriate computation methods have been developed for different systems of sampling.

The concept of incomplete penetrance

In contrast to the situations discussed above, in which the basic genetic system is uncomplicated, many human disorders with a specific genetic basis do not follow the well defined simple Mendelian patterns. Some traits show only a partial expression in the heterozygous gene carriers, just a fraction exhibiting the observable characteristic. One thus speaks of incomplete penetrance, described as the percentage of those possessing the appropriate genotype which actually develops the corresponding abnormality or phenotype. Dominance with incomplete penetrance is a frequent mode of transmission for human disorders, recognized more often in man than in experimental animals, because even minor human afflictions are likely to come to medical attention. The risk of such disorders in the relatives of index cases is proportional to the likelihood of possessing the mutant gene, and one does not expect to see the disease in all the gene carriers. Occasionally there are recognizable concomitants of the disease which appear in all the carriers, so that penetrance in regard to that characteristic is then complete, being somewhat lower with respect to actual abnormalities, and still more limited in terms of the true disease condition. The penetrance rate is thus a function of what is being observed.

Mutant genes involved in the transmission of traits following dominance with incomplete penetrance exist in most cases at sufficiently low frequencies to make the occurrence of the abnormal homozygote unimportant. However, when the mutant gene becomes more frequent, homozygous individuals are likely to be encountered in addition to the heterozygous ones, with the corresponding disease condition liable to be more severe. If all the homozygotes show a serious abnormality, one can in reality think of that condition as being recessively inherited, while the heterozygous state is then associated with a milder dominant disorder. In such a case the same gene is dominant in regard to one disorder and recessive with respect to the more severe disease. Sometimes, however, the disease of the homozygote overlaps with that of the heterozygote, which in turn may show a continuity with the normal state, and in that event separate diseases cannot be definitively identified with the different genotypes.

Overall utility of pedigree studies

While the principal subject under consideration is the transmission of intellectual variation, within the normal range, readers will note that most of the above illustrations center around hereditary diseases or disorders. This is in part because genetic studies by their nature are usually dependent on comparisons of groups, the genes becoming apparent only through measurable differences, often disease states. Codominant genes are also thought to be important in the production of lesser degrees of variation, but presently available methods generally leave them undetected and therefore essentially impossible to study in terms of genetic models.

Although such genes are not necessarily related to disorders, the basic principles of transmission are adequately illustrated by disease examples.

Pedigree studies of codominant genes may in some instances be possible, for example in situations like the transmission of the shape of the nose within certain families. Only time will tell whether similar approaches can be developed in relation to personality traits, even if it seems certain that codominant genes also play a significant role in that area.

Despite the problems inherent in the use of family data, pedigree analysis still has a definite place in human genetics, and inspection of family records often makes it possible to obtain important leads, which then can be subjected to more rigorous scrutiny. The presence or absence of one of the standard mechanisms can frequently be decided in this manner, particularly if the disorder in question is relatively rare and if most cases indeed are related to one specific mutant gene.

Besides their utility in revealing specific patterns of family transmission, pedigrees can also illustrate general trends, which may either support or refute a genetic theory. At times it has even been possible to demonstrate that a disorder segregates with systematic regularity into the various branches of a large pedigree, when the distribution in the family does not otherwise appear to fit one of the standard genetic patterns.

Some modern-day academicians tend to view askance any approach to science based on procedures seeming as primitive as an inspection of family pedigrees. Admittedly everyone is more impressed with intricate machinery, and the tendency is to assume that the man with the complicated equipment is bound to produce the important data. It should not be forgotten, however, that at times such beliefs on the part of university professors with advanced techniques and methodologies have led them astray. The academicians failed to see that simple methods, such as counting different-colored peas, could be more productive than refined microscopy, thus leaving to Gregor Mendel, the unrecognized monk, the discovery of the basic laws of heredity. No equipment outweighs the importance of a superior mind.

Pedigree studies may still appear unsophisticated to investigators with access to computers, but in reality many scientists have become enslaved by their complicated machines, feeling that they must keep them in operation because of the heavy financial investment already made. The scientist with few tools can often enjoy greater flexibility, unencumbered by such considerations. Einstein made his contributions with a pencil and paper, while his more "successful" peers were trapped in the intricacies of their equipment.

It is the belief of many geneticists that solutions to important genetic questions will still come from pedigree types of studies, and in particular this seems likely in the area of inheritance of human intelligence.

3
Quantitative Family Studies

While the original formulation of the principles of genetics was dependent on "unit characters," few human traits are as simple as the presence or absence of brown eye pigment. Most normal characteristics are complex, involving many genes acting in concert to produce the overall effect. But even in these situations a nonfunctional gene frequently manifests itself either in the heterozygous state or in a homozygous condition as a "unit character," although the corresponding normal state is not recognizable apart from the total appearance of health or normalcy of the involved structures.

In studies of simple normal characteristics, like brown eye color, or of discrete abnormal states associated with specific mutant genes, such as achondroplastic dwarfism, data can be collected for certain types of matings, determining what fraction of the offspring exhibits the trait in question. In this situation, all individuals can be classified as either possessing or lacking the trait, leading to specific percentage figures.

Metric characteristics

When it comes to investigations of the participation of genetic factors in the development of more complex normal characteristics, which often result from the interaction of many genes, the situation is quite different. Here the trait cannot be rated as present or absent, as all individuals possess the characteristic, such as body weight or intelligence, albeit to a variable degree. In this situation there is no way to establish the percentage of individuals possessing the trait; rather all persons can be measured, and there will be a distribution of figures. Normal traits of this kind are sometimes referred to as continuous, metric, or quantitative characteristics. To study the contribution of genetic factors to their development, one must compare the observed measurements among individuals or populations. Although there is no doubt that genetic factors operate on continuous characteristics, a separate set of methods is needed for their study.

While most metric characteristics appear to be normal traits, probably dependent on many genes, there also exist specific pathologic states which merge with the normal distribution, rather than manifesting themselves as discrete diseases readily distinguished from the normal range. This is particularly true in the case of dominant inheritance with incomplete penetrance, where only a fraction of the gene carriers is recognizably diseased, those affected also tending to show variable expressivity. These are in reality marginal conditions, in which one functional gene is almost sufficient for normal health, but not quite adequate in all circumstances, leaving some carriers of one mutant gene disease-prone.

Mental disorders illustrate medical conditions which merge with the normal state without a clear interruption. The view has been prevalent that this results from the nondiseased state simply covering the full range, starting in the region of excellent health and extending to the area of grossly disordered behavior. However, it would in reality be strange if no diseases existed in this area, so that it seems more likely that true abnormalities are indeed found at one extreme of the spectrum.

Multiple-gene systems

Metric characteristics, such as stature and skin color, are usually dependent on multiple genes. In studies of polygenic inheritance the standard methods employed are mainly statistical, the individual genes usually remaining unidentified. While a specific characterization of the involved genes would be much preferable, this may not be possible when only small additive effects to the observed trait are contributed by each member of a gene system, no methods being available for a separate detection of each one. Eventually chemical methods may be developed for measuring the action of the separate genes, but at this time all that can be observed is an overall product of all the member genes.

Some investigators have postulated that polygenic inheritance may be the cause of certain diseases, although at present such mechanisms are established to be involved only in the variation of normal characteristics, such as intelligence or skin color. With a polygenic system one expects to see a bell-shaped distribution curve for the characteristic in question. To account for the disease state, a threshold effect is sometimes proposed, all individuals located beyond a certain limit being seen as disease-prone. However, no specific disorders have been usefully explained by such a model.

Family statistics

Many personality traits have been noted to run in families, including antisocial behavior, neurotic tendencies, alcohol abuse, aggressiveness, and intelligence. In some instances the degree of resemblance in regard to such characteristics has been quantitatively established, for example between parents and offspring, among sibs, or even among more distant relatives. One must remember that the fraction of shared genes drops off quickly as the relationship becomes more distant. Monozygotic (identical) twins are the only individuals who possess all genes in common, dizygotic (fraternal) twins being no more related genetically than regular brothers or sisters. All first-degree relatives (parents, full sibs, children) share one-half of their genes with the index case, while second-degree relatives (grandparents, uncles-aunts, half-sibs, nephews-nieces, grandchildren) share just one-fourth. Third-degree relatives (great-uncles, first cousins, etc.) share only one-eighth. In genetic studies it is not very often worth the effort to assemble quantitative data going much beyond the first degree relatives.

To obtain a more meaningful comparison of various relatives in regard to a given trait than that obtained by qualitative estimates of family resemblance, it is, however, at times desirable to attempt in some fashion to make appropriate measurements and assess quantitatively the actual degree of similarity between all available groups. One way to express such data is to think of selected index cases or probands and assess the degree of similarity to them in regard to the trait for various types of relatives. Data are thus obtained listing the degree of resemblance or the percentage sharing of a trait, covering monozygotic cotwins, parents, sibs, children, uncles-aunts, nephews-nieces, cousins, and so forth. Such data can sometimes be very useful in the analysis of the likely genetic mechanism.

Quantitative studies of this sort may be unnecessary if one is dealing with a well-defined disease, which appears to follow a standard genetic mechanism. However, in the evaluation of more complex traits, with less clearcut modes of transmission, extensive family data may be indispensable. In cases of recessive inheritance this type of information may turn out to be of little value, except as a way to exclude other possibilities. On the other hand, if either dominant or polygenic transmission is suspected, it is very important to establish the proportional risk observed in different relatives. Unfortunately, the risk expectations are essentially identical with either of these last two mechanisms, i.e., the probability of receiving a dominant gene or, alternately, of acquiring the appropriate members of a polygenic system is in the order of 50 percent for first-degree relatives, 25 percent for second-degree relatives, etc. With incomplete expression the rates of the observed disease condition or phenotype drop, but they still remain proportionate. If a genetic condition can result from more than one independent cause, however, the ratios are altered, and then the relationships are unclear. The existence of a proportional relationship, when this can be demonstrated, gives further support to a genetic etiology, specifically suggesting either dominant or polygenic inheritance.

One problem in quantitative studies of different types of relatives is that available family members often fall into different age brackets, and corrections may then be necessary to bring the results to a common denominator. For example, comparison of old grandparents with young grandchildren may require special procedures, if the characteristic under study is age-dependent. At times the investigation can be designed in such a manner that this problem is bypassed.

Qualitative inspection of extensive pedigrees has relatively limited application in relation to frequent specific disorders or quantitative familial traits. Occasionally, however, such an approach may permit a definitive distinction between a polygenic mechanism and dominance with incomplete penetrance. A polygenic system is generally dependent on very frequent genes, this being mandatory for all the factors to occur together in the same individual with some regularity. Such genes must exist in all branches of a large pedigree, while a single gene associated with incomplete penetrance, being less common, is expected to show segregation into some branches, leaving others less affected. In this respect the two systems thus give different results, making a distinction possible when sufficient family data are on hand.

Mathematical relationships in family studies

Population genetics is a highly mathematical field, and various formulations and derivations have been published in the scientific literature. To avoid mistakes one must be aware of the distortions which occur in connection with data-collecting procedures. Precise statistical methods are also required to test whether the findings conform to the expectations based on different genetic models.

Often the data collection can be planned in such a manner that simple genetic ratios are to be expected. For example, with recessive inheritance the basic risks of the disorder in sibs of index cases are 25, 50, or 100 percent, depending on whether neither, one, or both parents are also affected. If index cases are randomly selected from a large population, the likelihood of a family being included becomes directly proportional to the number of affected sibs in the family. This results in an automatic correction for the loss from the material of families at risk, which by chance have no affected members, and one can therefore simply count all sibs, leaving out the index cases. However, if the study is designed in such a manner that all affected sibships in a given population are detected, the observed relationships are different, and other computation methods must be used.

An alternate approach to testing specific genetic hypotheses is to derive secondary formulae for the expected rates in various groups, basing these on different models and taking into account predicted distortions. The data are then compared with the new expectations to see which model gives the best fit. Computerized programs have been employed with this kind of approach, allowing for additional flexibility.

An important relationship in quantitative genetics is the Hardy-Weinberg law. This principle makes it possible to calculate the relative frequency of the three basic genotypes, when the rate of occurrence of the mutant form of a gene is known. Thus if p is the frequency of the native gene and q that of the mutant form, these being the only significant alleles, the relationship $p^2:2pq:q^2$ gives the relative frequencies of the $AA:Aa:aa$ genotypes, provided matings occur at random with respect to this gene and survival rates are equal. Since $p=1-q$, the expression can be modified to include just one variable, the frequency of the mutant gene being the only parameter that needs to be determined to make the computation possible. It should be emphasized that if there is preferential mating or differential fertility within the different genotypes, the basic relationships may be distorted.

In studies of metric traits it is often necessary to draw a somewhat arbitrary line between those accepted as "normal" and individuals classified as diseased. Different investigators are then unlikely to employ the same criteria, and one is faced with a considerable variation in the final figures. One useful device to circumvent this problem is to express all the results in comparative terms, for example in multiples of the disease rate in the overall population, if that information is available, based on the same method. Sometimes one particular group of relatives is encountered in abundance in all the studies, for example the sibs of index cases, and then the rate for that group can be used as a standard. Such procedures permit a more valid assessment of the results of different investigators in studies collating the overall data.

It cannot be overemphasized how necessary it is to be aware of the potential pitfalls and of the predictable distortions before one embarks on the collection of extensive data in quantitative genetics. Of course hindsight is always better than foresight, and it may be impossible in the beginning to tailor data collection procedures to the hypotheses that emerge as the work progresses.

In reality the mathematical concepts that need to be mastered are not very complex, and they can be handled by most investigators. The more important source of mistakes is the lack of awareness of the risks, with consequent failure to take them into account. This statement should not discourage clinicians from undertaking genetic investigations, since the truth of the matter is that they indeed have been the principal contributors to progress in human genetics. Laboratory geneticists, who often limit themselves to experimenting with flies or at best to the shuffling of data on blood samples which their technicians have rounded up from hospital laboratories, cannot possibly compete with astute clinicians, who constantly observe living people. The latter are obviously in a better position when it comes to gathering information requiring direct acquaintance with human traits. But clinicians should assess the situation carefully, before embarking on an extensive study, in order to avoid unnecessary mishaps. In the evaluation of data reported by others, one must also be on the lookout for systematic errors resulting from failures in the original planning.

4
Estimates of Heritability

People have always been aware that personality traits of parents are likely to reappear in their children. In recent years the view has been favored that some form of learning or environmental transmission is mainly responsible, but previously it had always been assumed that heredity was a major factor. If no increased family resemblance can be demonstrated with respect to a particular trait, hereditary factors need not be considered. For example, specific language characteristics are presumably mostly learned through environmental exposure, children showing little more resemblance to the parents than to other significant persons of the same culture.

Personality characteristics tend to be less clearly defined than many physical traits, but people still discern a resemblance, even in general attitudes, among genetically related persons. When father and son both show sociopathic tendencies, the public is inclined to assume that heredity is at least in part responsible. Such opinions, while of little theoretical value, may influence scientists to take a closer look and attempt to gather more definitive data.

Examples exist of disorders with a family concentration which in the past had been thought to be largely inherited, only to be later established as mainly environmental in origin. For example, leprosy was seen to run in families and was by many considered to be inherited until the leprosy bacillus was identified. A similar situation existed in regard to tuberculosis. However, the possibility was in reality never ruled out that inherited susceptibility may have been important in both these diseases. The lesson is obvious, nevertheless, that one must exercise care in concluding that heredity is the chief factor until other possibilities have been adequately explored.

There also exist resemblances which erroneously have been attributed to external experiences or learned behavior. For example, the opinion is widely held that eating habits are to a large extent learned, thus explaining why obese parents tend to produce fat children. In reality it is known that physiologic processes in the brain are important in the regulation of appetite and body weight, and this mechanism is so precise that an adult person tends to maintain a steady weight over long periods, the body mass only gradually increasing as middle age is approached. Adequate systematic studies are lacking, but there is much evidence that genetic factors play a major role in body configuration and weight control.

Occasionally, significant resemblance cannot be demonstrated between parents and offspring, even though one is dealing with a truly genetic condition. For instance, reports exist showing no correlation in regard to nearsightedness between Eskimo parents and their children. Since myopia is most likely a recessively inherited trait, rare in adult Eskimos, one would not expect such a resemblance, but correlation should occur between sibs, as the studies indeed demonstrate. Negative findings must thus also be interpreted with caution.

When a condition is thought to have a genetic basis, but pedigree studies and quantitative family investigations lead only to suggestive data, showing concentrations of cases in kinships, without establishing definitive Mendelian mathematical relationships, other methods become necessary to ascertain that genetic factors indeed are operative. Questions of this kind are particularly likely to arise in connection with personality traits, since one school of thought favors the view that mental characteristics may be mainly acquired through environmental contacts, while the opposing school emphasizes genetic origins.

For an assessment of the importance of genetic factors, without entering into the question of what the transmission mechanism may be, two principal methods have been developed, one using twin data and the other based on foster-reared individuals. Such information can be used to evaluate the relative quantitative contribution by genetic and environmental factors.

Twin studies

A powerful tool to evaluate the hereditary contribution to a human trait takes advantage of the existence of twins. Often such studies are restricted to same-sexed twins, comparing those that are monozygotic with a similar group of dizygotic pairs. While twin data are very useful in the assessment of genetic influences, they reveal essentially nothing about transmission mechanisms.

In white populations twins arise in a little over one percent of all births, approximately one-third of such twins being monozygotic, although that figure varies. Same-sexed twins are thus often approximately one-half monozygotic and one-half dizygotic. If the appearance of twins is so identical that they are easily mistaken for each other, the zygosity may be obvious, but sometimes monozgotic twins do not show that great a resemblance. In particular they may appear more dissimilar in the first year of life, so that even the parents may at that stage be unable to distinguish same-sexed fraternal twins from identical twins.

Various methods have been developed for a scientific determination of zygosity. At one time it was thought that monozygotic twins, who are derived from one fertilized egg, generally shared the same fetal membranes during intrauterine life, but this belief turned out to be erroneous. Fingerprints and other physical similarities have also been used. At present there is general agreement that the most reliable method for zygosity determination is a study of multiple blood groups, demonstrating that monozygotic pairs have all these in common. Unfortunately, there is considerable limitation in the availability of such data. First one must be able to obtain blood samples from both twins, and then have access to a laboratory equipped to do a thorough study. If only a few blood groups are determined, the results turn out to be no more reliable than the opinion of the twins and their families as to whether or not they are identical.

For genetic evaluations the most common approach in twin research is to assess the concordance rate in regard to the trait in question for each kind of twins. Another way of expressing this is to enumerate the proportion of the cotwins possessing the same characteristic as the index twins. In this type of study one is not measuring the precise degree of resemblance of the twins, but rather ascertaining the presence or absence of the trait in each member. The concordance rate is then expressed in percentage figures.

When the study is concerned with characteristics which cannot be rated as present or absent, including intelligence, body weight, or other metric traits, it is obviously not feasible to deal in concordance rates. In this situation a different approach is necessary, such as an estimation of the degree of similarity between the members of the various pairs. When a measure like the intelligence test score is known for one twin, the requirement is to determine how similar the score tends to be for the cotwin. From test data one can compute the degree of resemblance within pairs, assessing the findings separately for monozygotic and dizygotic twins.

In practice the latter types of study generally depend on the comparison of the correlation coefficients for monozygotic twins on the one hand and dizygotic twins on the other. In most instances it is possible to assume that same-sexed twins reared together share the same environment, so that greater similarities among monozygotic pairs reflect genetic identity. This reasoning has, however, been challenged in relation to personality traits, some investigators maintaining that monozygotic twins spend more time together and are treated more alike by their families.

The method used to secure the twin sample has an influence on the data, and the calculation methods must be selected accordingly. For example, if one locates all twins existing in a given population and studies their rate of concordance or degree of correlation for a given trait, the data differ from those obtained under conditions where a concordant or similar set is twice as likely to end up in the material. The latter would apply, for instance, if one locates the sample in a hospital or in a university, where only a fraction of individuals with the characteristic in question is likely to appear. The so-called proband versus pair methods are the main procedures applicable in these circumstances. In some studies, dealing with traits that are age-dependent, it may also be necessary to apply age corrections to the twin data, if the groups include pairs of different age levels.

Estimates of Heritability

Monozygotic twins generally show very high concordance rates or correlation coefficients with respect to physical characteristics, such as hair texture, eye color, or stature. Appearance of certain physical diseases has also been found to show high degrees of concordance. Mental traits show rather high correlation rates. In general, if concordance or correlation rates for a given characteristic are much higher in monozygotic than in dizygotic twins, this is seen as support for genetic factors. When concordance is complete in monozygotic twins, the trait is almost certain to be hereditary.

Although monozygotic twins share all their genes, it has recently been pointed out that they often differ in regard to brain laterality, this characteristic being influenced by early developmental events. If one member ends up with left-sided dominance, while the other happens to be right-sided, twins may consequently differ in mental traits without this being caused by the postnatal environment. More studies are needed to assess the importance of such factors, before it is concluded that differences between monozygotic twins must be attributed to childhood experiences.

Foster-reared persons

When other data, including twin studies, fail to settle the question of heredity, the ultimate approach is to study individuals reared away from the parents, in foster homes or orphanages. In ideal situations, such persons can be compared to a control group of individuals treated in the same way, but not related to index cases possessing the characteristics under study.

The time of separation from the biologic family sometimes becomes a major issue. In a retrospective study it may be hard to establish the exact time relationships, as records are likely to be scanty. The more the data are refined, rejecting cases for which the information is incomplete or those not separated from the family immediately after birth, the less total data one ends up with, resulting in greater statistical scatter.

Some environmentalists make the claim that intrauterine factors cannot be discounted in personality development, maintaining, for example, that a disturbed mother will adversely influence her unborn fetus. On the surface this claim seems unfounded, as even at the time of birth the process of myelinization is incomplete and many of the nerves still in an immature state. Although the brain is poised to respond to new stimuli and make rapid gains soon after emergence from the uterus, much of the normal brain substance is nonfunctional at that stage of development. A baby born with hydroanencephaly, almost the entire brain being replaced by fluid, therefore, behaves like a normal infant during the newborn period. These arguments, however, do not satisfy all the skeptics, and other data must be supplied. Few will deny, however, that a child whose contact with his father was limited to the moment of conception is unlikely to have derived environmental traits from him. Some studies have had to be refined to that point.

If one is dealing with somewhat rare disorders, data on individuals separated from their families may be hard to assemble. Even with more common conditions, sufficient material is not always accessible. For example, in some states laws make information on adopted children very difficult to obtain. Despite such adversities, dedicated investigators eventually seem to manage to supply the necessary data.

Various different designs have been employed in specific investigations of separated relatives. The most direct approach is to start with parents, frequently mothers, possessing the trait in question or known sufficiently well in regard to a quantitative characteristic under study, then locate children given up for adoption or placed in foster homes. Comparison of the children, respectively, with both the foster parents and the biologic parents makes the data more convincing than just a study of one set of parents. Sometimes it is necessary to accept foster parents of one group of children and biologic parents of another group. Alternately, the foster-reared children born to parents with the trait in question may be compared to a group of control children born to normal parents and reared under similar circumstances.

A different design involves affected index cases who were reared in foster homes, comparing their biologic relatives with the foster relatives in regard to the trait. In this approach it may be unwise to base the study on illegitimate children, since they are not likely to have full sibs, the group usually of greatest interest.

Often it is considered more valid to start with newborn children, following them as they grow up and thus performing a so-called prospective type of study. However, if one is mainly concerned with characteristics which cannot be evaluated before adult life, the delay may be undesirable, and many individuals can be lost from the study. In such cases there is in reality no objection to selecting individuals who were born some time ago, using recorded information which makes it possible to select index cases properly without knowledge about the outcome. One can thus avoid a long wait to complete the research, being able to match up the data on the current status of the offspring soon after selection of the study material has been completed. If properly carried out, this type of study is in reality a prospective type of investigation.

If it turns out that the study group indeed resembles the biologic relatives rather than the foster relatives, or, alternately, that the study population exhibits the characteristic in question to a significantly greater degree than a control group, it is hard to deny the involvement of genetic factors. At times, however, total isolation from the biologic relatives may be difficult to establish fully, and the criticisms may thus not be entirely quelled. Despite these problems, the study of foster-reared individuals is unquestionably a useful tool in establishing the operation of genetic factors in behavioral traits. As in other studies, proper age corrections may at times be necessary if the material is not uniform in that regard and the trait is age dependent.

Heritability quotients

In their quest for a quantitative expression, capable of describing the overall findings from various studies, geneticists have settled on the concept of "heritability," which is intended to be a measure of the fraction of a trait attributable to genetic influences. The basic concept has been applied in animal genetics as well as in the human field, different methods having been developed to meet specific requirements.

The degree of genetic influence can be computed from different types of data. If one is dealing with a trait suspected of being to a significant extent influenced by the family environment, heritability may be best assessed by data on individuals reared away from their biologic families. Correlation between various relatives, foster-reared or otherwise, can be converted into heritability estimates.

In regard to behavioral human traits, studies of twins have been found most useful, and many heritability estimates are now based on twin data. Mathematical formulae have been developed for conversion of measures on monozygotic and dizygotic twins into heritability quotients. Usually these estimates are based on the assumption that in both types of twins the members equally share the same environment. However, as was mentioned earlier, this approach has been criticized, the claim being made that monozygotic twins spend more time together and are treated more alike by their families. This makes it desirable to base the heritability estimate on twins reared apart, when such data are available.

Some heritability quotients have been computed from data on adopted children, comparing them with both the biologic and foster parents in terms of correlation coefficients, estimating from such information what fraction of the characteristic under study is produced by genetic factors.

Heritability estimates, even when arrived at under the most ideal circumstances, meeting all the criticisms that the skeptics can think up, are still not to be thought of as absolute values, reflecting a basic biologic process. The findings in this type of evaluation are dependent on the circumstances and are therefore subject to change. If the environmental conditions are altered, the fractional influence of genetic factors may no longer be the same. Genetic differences between populations can also result in different heritabilities being arrived at in different regions, the inherited variation accounting for less in a population which is genetically more homogeneous.

Despite their shortcomings, heritability estimates can be very useful as a way to summarize bulky data arrived at in different ways. When properly refined, they become the final common pathway to express complex findings from varied sources. Heritability figures have been criticized by many investigators, probably with some justification when they have been given too much credence by overenthusiastic advocates. But when they are used with appropriate caution, employing data gathered under favorable circumstances, there is no doubt that heritability quotients have a great deal of validity.

5.
Identification of Useful Mutant Genes

Even when a metric trait, demonstrated to be dependent on hereditary factors, is felt to have a polygenic basis, one can still entertain the notion that there may be a possibility of specifically identifying some of the individual component genes. As was mentioned earlier, the standard methods of dealing with multigenic inheritance simply confirm the presence of certain predicted characteristics of the distribution, without actually providing tools useful in the further pursuit of the genetic investigation.

Occasionally an abnormality associated with one of the genes in a polygenic system makes that gene individually recognizable. For example, while various genes in combination determine the height of an individual, the gene responsible for achondroplastic dwarfism causes not only reduced skeletal length, but also a very characteristic appearance. In the absence of the latter, the gene might just remain one of many obscure factors influencing overall stature.

In relation to personality traits new concepts have evolved in recent years, which promise to be of importance in the actual identification of specific genes. The basic ideas depend largely on considerations of gene frequencies.

Genetic polymorphisms

The designation "balanced polymorphism" has been applied to situations in which a mutant gene, identified in connection with abnormal development, exists at too high a rate to be explained by mutation pressure alone. The implication follows that such a mutant gene is probably in some way favorable to the organism, although that effect may for the time being remain unrecognized. The beneficial effect must somehow render the carrier of the mutant gene more competitive in the struggle for survival. Perhaps this effect can at times assert itself through a favorable personality structure, rather than through physical strength or increased resistance to diseases.

When a personality-related characteristic has been demonstrated to be inherited, it thus seems worth while to examine the group of genes responsible for various frequent disorders to assess whether possibly such genes may be involved also in the trait under study. The known number of frequent mutant genes is not large, but some have not as yet been definitively identified, so that the list may still grow.

Some of the common disorders thought to have a genetic basis are the following: diabetes mellitus, alcoholic tendency, schizophrenia, epilepsy, myopia, hypertension, gout, atopic allergy, baldness, and perhaps peptic ulcer, sociopathic tendency, and obesity. Certain of these abnormalities are known to be caused by single mutant genes. Discovery of an association of any of these conditions with a particular personality trait may enable one to identify a specific gene involved with both the disease and the behavioral characteristic. This approach opens a promising new avenue in the field of personality genetics.

Natural selection and gene frequencies

The majority of inherited human disorders are maintained at low frequencies by the forces of natural selection. A native gene designated A mutates to the nonfunctional counterpart a at a rate of close to 10^{-5}, i.e., one gene out of 100,000 mutating per generation so as to cause loss of metabolic activity. The abnormal form of the gene gradually builds up in frequency through mutation pressure, if there are no counteracting forces. On the other hand, if the heterozygous individual, Aa, is severely abnormal, perhaps to the point of

Identification of Useful Mutant Genes

not reproducing, the mutation may be quickly eliminated, affected individuals then remaining quite rare. This is thought to be the case, for example, with acrocephalosyndactyly, a presumably dominant condition leading uniformly to a grotesque appearance and mental deficiency. Dominant disorders with less drastic effects are, however, not weeded out as efficiently. Achondroplastic dwarfism is also caused by a dominantly transmitted mutant gene, but some affected individuals are able to reproduce, so that a fraction of the cases does not represent new mutations. On the other end of the spectrum, Huntington's chorea, a fatal neurologic disease with late onset, allows affected individuals to produce families before the illness becomes apparent, maintaining the gene in the population. While there is thus a variation in the frequencies of disorders following straightforward dominant inheritance, most are nonetheless kept at quite low rates because of constant negative selection. Usually the responsible genes exist at frequencies well below one percent.

Some mutant genes show no effect in the heterozygous carrier, while homozygotes, with two abnormal genes, are diseased and reproductively unfit. These are therefore purely recessive disorders. Since the usual gene mutations occur at fairly set rates, generally in the order of 10^{-4} to 10^{-5} per chromosomal locus per generation, one can calculate the frequency at which this type of gene should establish itself, when all homozygous individuals are eliminated from the reproductive pool. If the gene frequency, i.e. the fraction of all the genes at that locus which is of the mutant form, is designated q, the expression q^2 gives the rate of occurrence of recessive homozygotes. This is also the fraction of all persons who are eliminated by natural selection or, alternately, the fraction of the genes at this locus lost per generation. Once equilibrium is established, q^2 equals approximately 10^{-4}, i.e. the rate of elimination of abnormal genes equals the rate of origin of new mutant ones. From this q can be estimated as approximately 0.01. Many mutant genes which lead to severe abnormalities in the homozygous individual are therefore found to occur at frequencies of approximately 1 percent, the disease consequently appearing at the rate of one in ten thousand. If the abnormality is of lower severity, interfering less with reproduction, the gene frequency becomes somewhat higher, but still any gene with only detrimental effects remains infrequent. When a partial abnormality also exists in the heterozygote, perhaps not morphologically recognizable, this results in further selection against the gene.

Beneficial disease-related mutant genes

When an inherited abnormality is seen to exist at a much higher frequency than one would off-hand expect, geneticists conclude that the responsible gene must somehow produce a favorable effect, in addition to the more apparent detrimental condition. A classic example is the sickle-cell anemia gene, which reached extremely high rates in areas of Africa heavily infested with falciparum malaria. The explanation turned out to be that the heterozygous carrier was resistant to the disease, and the gene thus enabled man to survive in tropical regions which essentially were uninhabitable for normal humans. The price for survival was early death of the abnormal homozygotes, but basically the altered gene served a very useful function. As was stated above, geneticists refer to this kind of phenomenon as balanced polymorphism, meaning that the abnormal gene, because of its favorable effect, exists at a much higher rate than one would expect as a result of mutation pressure, counteracted by natural selection against the associated disease.

Balanced polymorphism can be viewed as a general principle, related to the concept sometimes referred to as hybrid vigor. Any mutant gene which exists at a high frequency may therefore be examined to assess whether it is involved in a demonstrably favorable influence, perhaps affecting a personality trait. The latter effect may then in reality be the principal action of such a gene, the disease condition also associated being a secondary effect. A search among frequent genes thus has a potential for identification of some of the specific personality-related genes which presumably do exist. This concept can be applied to such traits as intelligence, creativity, aggressiveness, or perseverance. These characteristics are likely to have a survival value, the responsible genes being therefore favored by the forces of natural selection.

The very existence of frequent disorders dependent on individual mutant genes in reality spells out a requirement for some type of beneficial associated effects, which should be amenable to scientific study. Since reproductive success is the only avenue through which the high gene frequency can be maintained, the obvious first step in a systematic study is to locate sectors of society which under normal competitive circumstances have exhibited elevated reproductive rates, the latter measured in numbers of surviving offspring. Having identified the successful segments, one can probably assume that they exhibit an increase in the diseases in question, and the need then is to determine in what way such groups are genetically better fit than others. If some of the frequent genes have indeed attained their high levels because of an intelligence effect, measures of brain performance should reveal increased abilities among reproductively fit persons. In this manner it may be possible to confirm that some intelligence-related genes are operative in polymorphic systems, but only further explorations will identify the specific genes.

Evolutionary expediencies

It should now be apparent that the principle of balanced polymorphism deals with the same concepts that were earlier evaluated in chapter 1, only viewing the effect from an opposite vantage point. In the former discussion it was postulated that some forms of evolutionary adaptation could be achieved by nonfunctional genes, this being a quick way to arrive at a specific downward adjustment in the rate of a chemical reaction. In the present chapter the phenomenon has been reintroduced in terms of the actual existence of genetic conditions dependent on frequently occurring mutant genes.

The increased frequency of the mutant gene in a polymorphic system is actually an indirect result of its favorable effect on a chemical reaction sequence. To further illustrate this concept, the assumption may be made that a series of chemical steps, each governed by an appropriate gene-mediated enzyme, is responsible for the production of a given essential metabolite. Perhaps changed conditions now have created a situation in which a reduction of the quantity of this metabolite can be beneficial to the organism. The direct evolutionary approach to the needed adjustment would be to alter slightly the gene governing the rate-limiting step, thus decelerating the overall reaction rate. Unfortunately, however, this type of adjustment is likely to require thousands of years. A quicker way of reducing the amount of product takes advantage of metabolically nonfunctional genes already existing in the gene pool, selecting out possibly another step in the reaction sequence, rather than that presently rate-limiting, to achieve the optimum overall rate. Only some individuals in the resultant population end up fitting the required type, but in a species able to take advantage of group adaptation this can often be sufficient. Grossly defective individuals fail to survive in the struggle for existence, but the organism can, at some sacrifice, achieve an immediate adaptation, instead of waiting for thousands of years for the more ideal solution. It does not seem unlikely that the very rapid evolutionary adjustment of human populations in recent times may indeed have utilized this biologic principle, whose availability is to a large degree dependent on the communication skills that man alone possesses.

6
Biochemistry of Gene Function

In the early part of this century, university students interested in biology were usually expected to have an exposure to genetics, being introduced, among other subjects, to the mode of inheritance of certain metabolic disorders in man. At the same time these students might attend a course in biochemistry, where the occurrence of inborn errors of metabolism was likely to receive cursory mention. The two instructors, however, probably were unaware of the connection between their fields, because the link between biochemistry and genetics had at that time not been established.

Today every student of biology learns about deoxyribonucleic acid (DNA) as the spiral-shaped macromolecule that forms the basis of the genetic material. The mode of its operation is explained in minute detail, demonstrating how the triplet code of DNA is translated, via ribonucleic acid (RNA), to spell out the amino acid arrangement of enzymes and structural proteins. Much of this knowledge became available through the efforts of microbiologists, who found bacteria and viruses to be especially suited to controlled manipulations and chemical analysis. Once the basic principles had been established, it was relatively easy to verify that the same phenomena were also operative in higher organisms, including man.

Methods employed in biochemical genetics

Many chemical abnormalities have been discovered in the course of other studies or even just by accident. No overall procedures exist which can be used to lead one to the chemical abnormality associated with a given genetic disease. Searches for metabolic disorders involve trial and error or educated guesses, although refined methods are available to investigate whether certain classes of chemicals perhaps are involved. For example, one can start with some of the body fluids from a person with a hereditary disease and search for abnormalities in the amino acids by the use of refined chromatographic techniques. When one also takes advantage of the possibility of labeling certain metabolites with radioactive isotopes, in order to ascertain their fate, or of the use of spectrophotometric determinations, which are capable of detecting very minute amounts of chemicals, the possibilities of success are further enhanced. Bioassay methods, making use of biologic preparations capable of responding to very small quantities of active metabolites, allow still more refinements.

Often chemical abnormalities have first been explained in lower organisms, the knowledge being later applied to human systems. Bacteria can be exposed to radiation to produce mutations which interfere with the synthesis of essential metabolites. Mutations are actually random events, potentially affecting any gene, but from a mixture of mutants one can choose for study those involving the chemicals under investigation. When specific biochemical abnormalities have been elucidated in such systems, they may turn out to have counterparts in human disorders, which then can be explored further. The use of induced mutations in lower organisms has thus become one of the tools for studies of human abnormalities, because all organisms are much alike in basic biochemical machinery.

Even with all the above methods, a biochemist may still be at a loss about how to approach the problem of finding the chemical aberration associated with a specific genetic disease. This is why most hereditary disorders are still unexplained in terms of the biochemical disturbance. Sometimes the abnormality may also be a localized one, so that it can be found only if the search is directed toward the appropriate anatomic region or tissue.

Knowledge about the operation of the normally functioning

gene is also frequently arrived at by a study of genetic abnormalities. A chemical reaction sequence may ordinarily proceed from a starting substance to a distant end product, without any evidence being seen of the intermediates, which usually exist only in a small quantity. A genetic abnormality, leading to the arrest of the process at one of the steps, may then reveal that a certain metabolite is a member of the reaction chain. By studies of this kind, coupled with radioactive tracer methods and inhibitor studies, biochemists have gradually come to know a great deal about the normal chemical operations which take place in the very complicated human machine. However, much still remains to be learned, especially about the biochemistry of the central nervous system, which in the living state is very inaccessible to chemical determinations or metabolic studies. In particular this applies to the neurohormones, which often act locally and involve only minute amounts of very active substances.

Inborn biochemical abnormalities in man

Biochemists have established thoroughly how the transformation of chemicals in the body occurs in a step-wise fashion, each reaction being guided by an appropriate enzyme, which in turn is under genetic control. These systems operate for the synthesis of complex chemicals, needed for the body machinery. They also engineer the breakdown of waste products or the oxidation of substances utilized as sources of energy. In his search for metabolic abnormalities a biochemist thinks in accordance with this overall model.

In attempting to explain the biochemical mechanism of any genetic condition, it is important to keep in mind that the most frequent type of expression of a genetic abnormality is a partial or complete arrest of a chemical step in a reaction sequence. One often hears alternate proposals, for example that a mutant gene may lead to malfunctioning of the receptor of a nerve impulse or block the action of a hormone through interference with its normal site of action. A more direct mechanism of the usual type should, however, be given prime consideration, before one embarks on a search for an unusual type of expression, although the latter possibility certainly should not be overlooked. It is also useful to keep in mind that often a pharmacologic agent will essentially duplicate a genetic abnormality, partially or totally blocking a particular chemical step by interfering with the operation of an enzyme. This creates the possibility of finding a pharmacologic model which resembles the genetic disease.

Certain familiar diseases are frequently utilized to illustrate how inherited enzyme deficiencies manifest themselves. One such disease is phenylketonuria, a disorder leading to severe mental defect in individuals who are homozygous for a specific mutant gene, consequently lacking the corresponding enzyme. The normal allele, either in single or double dose, guides the formation in the liver of an enzyme which converts the amino acid phenylalanine to tyrosine, the latter then being further metabolized by other enzymes. While phenylalanine is an essential amino acid, present in most proteins, a regular diet contains more than the necessary amount, and too much is toxic if the breakdown mechanism is not working. The toxicity of excess phenylalanine in phenylketonuric individuals can now be largely averted by a special diet, low in the offending amino acid, provided the disease is diagnosed in infancy.

Many other recessively inherited diseases follow this prototype, the abnormal homozygote lacking an enzyme necessary for one step in a chemical sequence. The actual disease may result from a deficiency of a specific chemical, from an accumulation of a toxic intermediate, or from abnormalities in development resulting from the metabolic imbalance.

Dominantly transmitted disorders have been more difficult to study chemically, presumably because the corresponding biochemical reaction is only partially blocked, making it harder to demonstrate a difference between affected and unaffected individuals. An absence of an enzyme is obviously more readily detected than a somewhat reduced amount. Since metabolic adjustments also occur in response to altered balances, it may be difficult to ascertain whether an observed quantitative difference is a direct result of the genetic condition.

Although partially dominant disorders are very frequent in man, few of these have been adequately explained in biochemical terms. Limited chemical knowledge is available about such conditions as gout, where excessive uric acid is the offending metabolite, but the pathology of most such diseases is far from being fully understood. Acute intermittent porphyria has been shown to be associated with a disorder of a specific enzyme, but here, too, it is uncer-

Biochemistry of Gene Function

tain whether this is the primary change. In general, chemical studies of dominant disorders are in their infancy, and one can expect important progress in that area in the future. Presumably such disorders are usually associated with changes in reaction rates, which are difficult to investigate, as large variations occur normally in the activities of various enzymes.

Metabolic activity of mutant genes

Since there is not universal support for the view expressed earlier, that disease-related genes are in the majority of instances of the nonfunctional variety, further discussion of this concept seems in order at this point. Many geneticists prefer to view all mutant genes as simply altered variants of the native or primeval allele, assuming that any form of the gene can potentially serve a metabolic function. It cannot be disputed that many examples indeed exist where metabolic activity of a mutant allele can be demonstrated. For instance, the gene associated with sickle-cell anemia has been found to differ only in one DNA base substitution from the native gene, and this results in the formation of a protein in which one amino acid is replaced by another, metabolic activity still being retained. However, it should be pointed out that this very specific mutation is essentially nonrecurrent, the rate of its origin being so low that it apparently arose in one place and is not found in most populations.

When one examines more commonly occurring mutant genes, such as that found in phenylketonuria, recurrent mutations are encountered in all populations, indicating a much higher rate of origin. Furthermore, since all phenylketonuria genes are for clinical purposes alike, even though of multiple origin, and all lead to failure to form the corresponding enzyme, the only logical explanation appears to be that all such genes share the property of inactivity, whatever the basis of that may be in any particular case. Most genes responsible for recessive disorders, which to this point have been explained in biochemical terms, are associated with failure to form the corresponding enzyme. The hemoglobin disorders, including sickle-cell anemia, are an exception, and they do not involve enzymes in the usual sense. These can actually be viewed as being related to codominant genes, and it just happens that individuals homozygous for the sickle-cell allele suffer from the adverse physical properties of the specific form of the protein, although their hemoglobin can still carry oxygen.

The concept that the ordinarily encountered mutant genes are mostly metabolically nonfunctional was originally formulated in connection with inborn errors of human metabolism. Subsequently it was postulated that most of the mutants studied in fruit flies were similarly inactive, since a known total absence of the gene led to the same expression as the abnormal homozygous state. Later this concept was expanded greatly in connection with investigations in biochemical genetics of microbes. However, since the latter studies were tailored to select out defective metabolic mutants, the question could be raised whether the findings in that material are fully representative. Indeed, it has been amply demonstrated that it is possible to establish a graded series of mutations with different degrees of activity, and no doubt the variants showing lesser degrees of metabolic interference were in fact excluded from the earlier microbial studies.

Although on theoretical grounds one can thus postulate that metabolic inactivity is not at all mandatory for most mutations, a difficulty quickly arises with this concept in relation to human disorders. It remains a fact that, in general, hereditary diseases are found to be equivalent in different families. Abnormal homozygotes which have been carefully studied have also in the majority of instances turned out to fail to form the corresponding enzyme. There is no way to adequately explain disorders like alkaptonuria, phenylketonuria, or galactosemia without assuming that the mutant gene is an inactive one in all carrier families. Alleles of different degrees of inactivity would produce an array of disorders. This is why it is proposed here that disease-related genes are mostly nonfunctional, while less drastically altered variants of the native gene, still possessing metabolic activity, presumably constitute mostly the codominant series of genes. Admittedly, it seems established that certain partially active mutant genes can be responsible for some diseases. Nevertheless, one would be hard-pressed to come up with an explanation of the generally encountered equivalence of disease-related genes without involving the hypothesis that such genes are in the majority of instances metabolically inactive and therefore identical. This concept explains many facts and appears to have much validity.

It will consequently be assumed in future chapters that the abnormal genes encountered are mostly of the inactive variety, regardless of whether the corresponding disease is classified as dominant or recessive. This assumption imparts a unifying character to the disorders under consideration, and in the study of transmission mechanisms, involving numerous different families, it is desirable to postulate a certain degree of equivalence of the proposed genes. If this assumption results in data which can be interpreted in terms of standard genetic systems, that in itself gives support to the soundness of the approach. It should be emphasized, however, that this conception, although preferred by the author, is not a necessary tenet for the validity of the genetic arguments.

SECTION II
Genetic Transmission of Giftedness

7
Definitions of Intelligence

All healthy human beings are in possession of certain fundamental abilities, which seem to be built into the nervous system of the normally developed organism. Similar basic behaviors are in fact seen also in various animals, so that even a chicken recognizes its mother, learns the location of food and water, finds its shelter, and participates in group behavior. When sickness interferes, some of these functions may no longer be seen.

At the lower end of the human intelligence spectrum, one is likely to find an admixture of both the less clever persons and those who carry some disabling abnormalities. It obviously is not realistic to expect equivalent behavior in a sick individual with good potentials and in a healthy person with limited endowment, although often such individuals are for reasons of expediency grouped together. Psychologists have in fact observed that pupils exhibiting relatively low scholastic performance can often be separated into those who outside the classroom behave like other children, having the abilities to recognize their friends or participate in games, and those who are somewhat limited in all functions, being clumsy, poorly coordinated, and unable to reciprocate in play. The first group may simply be dull, the second perhaps diseased. If the dysfunction is a severe one, a pathologic state is likely to be recognized.

GENETIC TRANSMISSION OF GIFTEDNESS

This book deals mostly with differences in intellectual endowment among healthy persons. Attempts will be made to unravel what factors elevate certain individuals above the limited "normal" range, hoping to identify some of the genetic influences. Although the subject of the present section is designated "giftedness," it deals largely with concepts which often are termed "intelligence," particularly in connection with school assessments.

The general concept of intelligence, as commonly used by the public, actually embraces all areas of learning, reasoning, and formulating new ideas. It is in this broad sense that the term will be used in this book. The different aspects of intelligence can vary more or less independently, and thus memory, judgment, alertness, or creativity can be subjected to separate studies.

Intelligence tests

Psychologists sometimes employ the term "intelligence" in a more restricted sense, particularly in studies limited to humans. The extensive and relatively successful application of intelligence tests has provided a large body of information, which makes possible evaluations of various groups and comparisons of different populations. The enthusiastic acceptance of these procedures has led to new definitions, and many psychologists now describe human intelligence as those properties that intelligence tests measure. While this involves a somewhat circular reasoning, the approach nevertheless has applicability for those doing research based largely on test data.

Intelligence tests grew out of school needs, being originally developed as a tool to identify students with either limited or superior potentials. It is therefore not surprising that they correlate best with school achievement. The scores on such tests tend to remain relatively stable for a given individual over many years, and they are indeed a good predictor of academic learning ability. Correlation with occupational status and social success is also reasonably satisfactory.

Many different tests have been designed, some meant to be administered individually by a psychologist, others tailored for group use, some quite dependent on language skills, others designed to avoid word usage. Each test is multifaceted, some aspects measuring ability for immediate recall of unfamiliar items, others testing funds of knowledge, still others involving manipulation of concepts. Obviously it is to some extent arbitrary what stress is placed on the different aspects of such tests, but trial and error have served to guide the selection of various items to be included.

The scoring of intelligence tests is done in terms of IQ points. Originally the estimated mental age, converted into a percentage of chronologic age, formed the basis of the intelligence quotient, but this method is no longer in use. The current approach is to define the average score for individuals of any given age as 100 and only administer such tests as are appropriate for that age level. The overall range of scores for the population at any age follows a normal distribution curve, this again being to some extent by choice, as the structure of the tests is adjusted to yield such a pattern. The standard deviation of intelligence tests is usually 15 IQ points, meaning that two-thirds of all individuals fall between IQ 85 and 115 or 95 percent between IQ 70 and 130. One-half of all people are between IQ 90 and 110.

Individuals below IQ 70 are considered mentally defective, those above IQ 130 of superior intelligence. Because of slight irregularities in the curves, the 5 percent of the population found outside these limits partitions into approximately 3 percent in the defective range and less than 2 percent in the gifted category. Only one in three hundred achieves a score above 140, although these figures vary, depending on the specific test or the population studied.

Significance of IQ scores

At present there is much discussion whether IQ scores are truly indicative of intelligence, as the concept is used by the public. Especially heated is the debate over the reliability of the tests in the comparison of different ethnic or cultural groups. This subject will not be dealt with here, as it is not relevant to the issues at hand, and adequate coverage is available elsewhere.

It is not disputed that the tests have value as predictors of school performance and later occupational adjustment within a cultural group, and they do provide the most comparable data available for studies within such populations. Whether they measure all of "intelligence," or just certain aspects, is another matter, and this is why such designations as "academic intelligence" or "giftedness" seem useful. There is abundant evidence that creativity is not measured by these tests, so that persons found to achieve very high IQ scores are not necessarily likely to be creative.

Some attempts have been made to design tests which would assess on the one hand factors like short-term memory or retention and on the other reasoning ability or judgment. All of the standard IQ tests contain the latter element, sometimes referred to as the *g* factor or general intelligence, although it is more concentrated in some tests than in others. At this time no tests are in general use which only measure short-term retention without touching on reasoning capacity.

Alternate types of mental performance

Other kinds of mental abilities obviously exist besides those involved in academic or test performance. For instance, it is well known that some persons who achieve poorly in school later turn out to be effective members of society, even at times its leaders. One such example is Winston Churchill, who was considered retarded by his teachers.

No specific tests are available which discriminate between low IQ pupils with hidden potentials and others who remain genuine failures. Intuitively, however, the peers of individuals destined to be successful are aware of their abilities, even if the tests or school ratings fail to measure them. It is not uncommon for those dealing with children with behavior problems to see a discrepancy between assigned IQ scores and skills exhibited in play or other real-life settings. Abilities in social judgment, cleverness in trading, or skills for manipulations of people may be critical attributes which are inadequately measured by scholastic types of tests, although such characteristics obviously have value in subsequent life situations.

On the other side of this issue, everyone knows individuals who can adequately master school requirements, exhibit social grace, or keep track of friends and current events, but display appalling performance in regard to sound judgment, basic understanding, or even meaningful interpretation of established facts. Again the IQ tests may fail to distinguish between such persons and those who can learn and also consistently show good judgment.

Some scholars discuss the concept of intelligence in ways that differ radically from the approaches encountered in the psychologic literature on IQ measures. Sometimes true human intelligence is even described as the ability to invent problems, or in more specific terms as the knack for identifying soluble problems. Intelligent activity in this sense is not necessarily correlated with scholastic ability. Animals are able to learn, but conceptual thinking and in particular the capacity to abstract is seen by many as the very essence of human intelligence, distinguishing it from behavior shown by other organisms.

Culture forms a very important part of human evolutionary gains, and while the development of habits and traditions can be viewed as a nongenetic form of adaptation, which has enhanced man's capacity to make rapid progress without waiting for slow-moving biologic changes, the basic ideas have had their origin in the minds of a few individuals, whose superiority is dependent on genetic factors. Culture operates through language and is dependent on cognitive ability and symbolic thought. There is no real evidence that the factors which allow for the development of human culture, in turn leading to civilization and industrialization, are adequately measured by IQ tests. Other kinds of abilities may be equally important in the realm of human progress. Individuals who fail in school but perform brilliantly in social leadership obviously are capable of assembling facts and manipulating information, and in that sense they are gifted. Such factors as restlessness or increased anxiety can, however, interfere with scholastic performance.

It is thus apparent that there exists no overall agreement on a definition of true intelligence. For the purposes of the present discussion the designation "giftedness" will be equated largely with IQ scores or school achievement and to a lesser extent with social standing or occupational success, because data on the former are readily available. It would be possible, as an alternative, to emphasize specific abilities, including mathematical skills, but there is even less agreement that such measures reflect true intelligence or superior judgment. Some educators prefer to consider IQ scores to be indicators only of general scholastic aptitude, avoiding the word intelligence in reference to test results. Such caution seems unnecessary, as long as one maintains awareness of the limitations of the concepts and of alternate interpretations.

8
Hereditary Contributions to Learning Ability

Students of the family aspects of intelligence have utilized all the approaches to genetic evaluations discussed in section I. Pedigree analysis is in this instance relatively noncontributory, since it is difficult to map out families in terms of intellectual endowments. Quantitative family studies, twin investigations, and comparisons of foster-reared individuals have each contributed impressive data. The overall result, as will be discussed in some detail below, is a substantial support of genetic influences.

Opposition to genetic theories

Notwithstanding very convincing scientific evidence, many educators have been reluctant to acknowledge that differences in intellectual endowment may be to a large extent traceable to hereditary factors. Psychologists have also tended to emphasize environmental interpretations, supporting the concept that exposure within the family to scholarly experiences should be viewed as the crucial factor in the development of learning inclinations and intellectual curiosity. It has, furthermore, been considered a democratic virtue to assume that all persons are born biologically equal, even politicians having found it expedient to claim that early educational or even nutritional deprivation is perhaps the principal reason for later impaired performance by some groups.

While there obviously can be no disagreement that the operation of remediable deficiencies must be given proper consideration, it is in the end useless to let emotional preferences supercede scientific evidence. Environmental theories must be subjected to the same rigorous testing that generally is demanded for genetic explanations, no justification being possible for continued adherence to unproven hypotheses, if the scientific data point in a different direction. Even if it may for the moment seem kind or humane to pretend that observable constitutional differences do not exist, no one is in the long run done a favor by denial of the scientific facts. Programs based on deliberate disregard of the actual data are doomed to failure anyhow and should be abandoned. A heavy price may have to be paid by future generations if important national policies end up being adjusted to temporarily expedient misconceptions.

Family resemblance in learning aptitudes

It has always been known that intelligent parents give rise to bright children, and the public has generally interpreted this as a sign of hereditary differences. This is certainly not equivalent to suggesting that learning is of limited benefit. The impression has, however, been prevalent that some individuals are born with superior mental powers, while others are from the start deprived, no amount of schooling being capable of making up the difference.

As an example of the family concentration of general scholastic ability, the data in Table 8-1 illustrate how top scholars in Iceland tend to be related to each other. The figures are derived from the college preparatory school in Reykjavik, dealing with male relatives of those who graduated at the top of their classes during the last century. For most of the period covered this was the only such school in Iceland. One can estimate that close to one thousand surviving males were born in that country each year, and empirically it turns out that the rate of top graduates among several thousand randomly chosen individuals agrees reasonably well with that figure.

The probability of being Iceland's most successful graduate is greatly increased for those who are in some way related to a top graduate. The figures are too small for an exact estimate, but it appears that a first-degree relative of a top performer has a fiftyfold elevation of the probability of also achieving that status. A definite increase is also seen in more distant relatives. Although some differences in opportunity for schooling existed in the first half of the period, Iceland has been an essentially classless society, so that bright students from poor families have not been deprived of the opportunity to attend school.

Table 8-1
Likelihood of graduating at the top of the class from the college preparatory school in Reykjavik during the period 1871-1960, comparing male relatives of top graduates with the general population

Relationship	Number studied	Top graduates per thousand
First-degree relatives	227	32.4
Second-degree relatives	420	9.5
General population	12,266	0.7

Studies were done in several countries many years ago on the resemblance of orphanage-reared children to their relatives. It turned out that a child's mental ability was highly correlated with the parents' occupational levels, even though contact with the family had been minimal.

With the emergence of IQ test results, more definitive studies became possible, and specific data have been gathered on the degree of resemblance between different groups of relatives. On the basis of genetic relatedness, one would predict that correlation in IQ between first-degree relatives should approximate 50 percent, if genetic factors indeed are the principal determinants of measured intelligence. Since the observed figures actually are found to be almost precisely as predicted, it is hard to understand how some investigators feel justified in maintaining that this gives no support to the genetic hypothesis. After all, it is the essence of science to make predictions and compare them with the observed data.

If one expresses the measured values for various groups of individuals as the similarity of their IQ scores to those of index cases, monozygotic cotwins, who share all their genes with the probands, attain almost identical scores. All first-degree relatives, sharing one-half of their genes with the probands, are found to show approximately 50 percent correlation, while more distant relatives exhibit less similarity to the index cases.

Additional studies of the IQ test resemblance within pairs of monozygotic twins have indicated that the results are so similar that the score achieved by the cotwin is in most cases essentially what one might obtain by simply retesting the proband twin. The reported correlations range from 87 to 97 percent, depending on how much correction is applied to the data. While twins reared apart exhibit somewhat reduced correlation coefficients, the values vary from 75 to 89 percent. The 122 reported pairs of separately reared monozygotic twins thus resemble each other more closely in IQ performance than any less-related individuals. On the other hand, dizygotic twins, whether reared together or apart, are no more alike than regular sibs, their correlation figures being in the order of 50 percent.

IQ of foster-reared children

Both the correlation studies of various relatives and the concordance determinations in twins give rather convincing support to genetic factors in intelligence. However, even more impressive data have been reported on the comparison of foster-reared or legally adopted children with their biologic versus foster relatives. The total evidence from such studies has recently been reviewed by Munsinger.

Many authors have gathered data of this kind during the last half century, using various designs and applying different controls. Most of the studies are in good agreement with each other, generally indicating close resemblance of foster children to their biologic parents in IQ estimates, while little likeness is found between the child and his foster parents. The more complete studies include sufficient data to permit different forms of evaluations. For example, Honzik analyzed data reported by Skodak and Skeels, supplemented by her own study, and prepared tables and illustrations comparing the correlation on successive tests between adopted children and their biologic versus foster parents. At age two years, little correlation is apparent with either set of parents, since no satisfactory IQ tests are available for that age group. On successive tests the correlation with the biologic parents rises to the same level as is seen between parents and their home-reared children, but the correlation with the foster parents remains insignificant. Some of the data are shown in Table 8-2. The approximately 35 percent correlation between parent and child seen in these studies is lower than the 50 percent correlation on IQ tests reported for first-degree relatives, because the information on the parent groups is limited to their education attainment. When the latter is compared to the children's IQ, the measures usually show somewhat less definite relationships than actual IQ scores for both groups. For comparison with the Skodak and Skeels study, Honzik gathered data on children reared by their own parents, again finding the correlation between the child's IQ and the parents education to be in the order of 35 percent.

Table 8-2

Correlation at different ages between the IQ performance of adopted children and the education levels of their biologic versus foster parents—data of Skodak and Skeels

Age of child (years)	Correlation coefficients			
	Biologic parents		Foster parents	
	Father	Mother	Father	Mother
2	0.03	0.04	0.05	—0.03
4	0.36	0.31	0.03	0.04
7	0.28	0.37	0.03	0.10
13	0.42	0.32	0.00	0.04

Munsinger lists many other studies of foster-reared children and makes an overall assessment of the available data, arriving at the conclusion that the observed correlations between different groups are almost those predicted on the basis of a purely genetic determination of intelligence. On the other hand, a comparison of the data with predictions from an environmental model shows great discrepancies.

The total information from various types of studies thus gives strong support to the hypothesis that intelligence is in large measure dependent on hereditary factors. Attempts to convert the findings into heritability quotients have consistently led to values of 60 to 80 percent, the contribution of environmental factors to measured IQ then accounting for 20 to 40 percent of the total variation in the societies studied. One can conclude that the evidence for hereditary determination of learning ability has become very convincing.

Nonacademic forms of mental ability

Because of the ready availability of IQ data in the form of numerical measurements, which lend themselves well to statistical manipulations, the above discussion has been based largely on academic performance.

There is, however, a widely held opinion that other intellectual traits, separate from those that enter into IQ measurements, are also influenced by genetic factors, although precise data, allowing for a truly scientific assessment, are scanty. Hopefully it will in the future become possible to supplement the IQ data or academic evaluations with quantitative measures of other personality variables, which also are important in the overall adaptation of an individual to his environment and in his advancement toward social success. There are indications from family studies that genetic factors play a role in qualities such as assertiveness, leadership, or social awareness, which often are seen in successful persons who failed on academic measures. Limited data on twins similarly suggest significant heritability of traits like extraversion, boldness, self-confidence, or sociability. Longitudinal studies of young children have shown that personality characteristics make their appearance early in life and remain relatively fixed, bolstering still further the concept of innate differences in behavioral attitudes.

Relatively nonspecific evaluations of the distribution of social success suggest intrafamily transmission of general mental abilities, basically comparable to the findings from IQ data. A study of the familial aspects of achievement was published by Galton a century ago, supporting a genetic tendency.

As an example of this type of investigation, the data in Table 8-3 illustrate the family concentration of listings in *Who's Who in Iceland*. The index cases are themselves listed, being otherwise randomly selected males born in the intervals 1851-1880 and 1881-1910. Their male relatives, born during the second interval and surviving past age fifteen years, are included in the study. It turns out that the first-degree relatives are three times as likely to be listed as the general population. Whatever the qualities may be that are being measured in this type of comparison, it is evident that a family trend exists, although this study does not establish a genetic cause. The figures are not a direct function of the degree of relationship, but if the value for the general population is subtracted from each score, the remaining figures agree reasonably well with a proportionate rate.

Table 8-3

Rates of listing in *Who's Who in Iceland* for various groups of males born in the interval 1881-1910 and related as indicated to index cases listed in the same source

Relation to index case	Number studied	Rate of listing
Brothers	470	22.8
Sons	368	18.0
Nephews	285	14.4
First cousins	489	11.8
General population	3456	7.0

Although studies of the heritability of scholastic success have yielded the most definitive data, there exists thus some support for genetic influences in other areas of personality development.

Genetic mechanisms in intelligence

Acknowledging that genetic factors appear to be operative, one can now attempt to evaluate the nature of the transmission systems. In view of the complexity of human intellect as a concept, it is not likely that any simple genetic mechanism can account for all differences between various individuals or populations. Intelligence is obviously not something that either is present or absent, but rather there is a large and continuous range of abilities.

Since scientists are in agreement that many separate genetic factors probably influence intelligence, sometimes positively and at other times negatively, some form of polygenic inheritance is considered the most plausible system. The observed family patterns indeed support this opinion.

The data reproduced in Table 8-4 summarize a large body of information on family relationships in regard to intelligence measures, which have been culled from the literature by several investigators. The figures given in the first column are the correlation coefficients between the individuals listed and the corresponding index cases, in terms of measured IQ. Most of these figures are based on many separate studies, each involving a large group of individuals.

Table 8-4

Summary of studies of correlation coefficients for IQ measures, comparing various types of relatives

Relation to index case	Degree of resemblance to index case	Fraction of genes shared with index case
Monozygotic cotwin, together	0.87	1.00
Monozygotic cotwin, apart	0.75	1.00
Dizygotic cotwin, same sex	0.56	0.50
Dizygotic cotwin, opposite sex	0.49	0.50
Sib, together	0.55	0.50
Sib, apart	0.47	0.50
Parent	0.50	0.50
Child	0.50	0.50
Grandparent	0.27	0.25
Grandchild	0.27	0.25
Uncle-aunt	0.34	0.25
Nephew-niece	0.34	0.25
First cousin	0.26	0.12
Second cousin	0.16	0.03
Unrelated, together	0.24	0.00
Unrelated, apart	—0.01	0.00

When one compares the correlation figures with the average fraction of genes shared with the index case by each relative, listed in the second column, the similarity of the numbers is quite striking. The size of the correlation for each relative indeed corresponds rather closely with the degree of genetic relatedness.

The best explanation for these findings is that intelligence in fact follows a polygenic system of inheritance, a mechanism expected to yield a normal bell-shaped distribution of scores for the overall population and also lead to a correlation in different relatives proportional to the degree of relatedness.

In reality, however, this conclusion has limited value in the sense of giving direction to further investigations. It offers little that is new, and really just reiterates what people have considered to be true for a long time. Nevertheless, the data place such opinions on a firm scientific foundation, narrowing the range of possibilities, defining the problem more clearly, and giving impetus to attempts to develop new avenues of approach.

Relation of intelligence to frequent disorders

In attempts at defining the direction of further studies, the new concepts of balanced genetic polymorphism turn out to be more productive. It can be assumed that improved intelligence has in the past had a survival value, and this assumption is in fact supported by studies which have indicated that persons of higher intelligence have tended to produce a larger number of descendants. Investigations reported on Asian populations many years ago established that at that time parents of higher achievement were likely to successfully rear an increased number of children. Although in Western countries there were at the same time reports on the largest families often including groups of low intelligence, more specific data suggested that only a fraction of dull persons ever married, so that the total number of offspring was still small for that segment of the population. It is, however, possible that these relationships may be changing at present.

As was discussed earlier, it actually seems probable that some of the genes involved in intelligence may be of the codominant variety, being genes which differ only slightly from the basic gene, both forms having qualitatively the same metabolic function. Unfortunately, however, there are no methods available at present which permit identification of superior codominant genes. The claim that some populations may have become homozygous for somewhat improved specific genes remains to be verified.

A different line of reasoning leads one to think of metabolically nonfunctional mutant genes with a polymorphic distribution. Since the polymorphism presumably results in most cases from the heterozygote possessing some unexplained advantages, it is always possible that in some instances the benefit may be an intellectual one. Geneticists refer to independent effects of the same gene as pleiotropy, and many genes are known to influence more than one physiologic system. A search for an intelligence effect among genes associated with frequent disorders thus seems to be in order.

One gene that comes to mind as a possible candidate is that

thought to be responsible for nearsightedness. Could it be that myopia has reached its high frequency in some populations because it leads to an enhancement of intelligence, aside from its effect on the anatomy of the eyeball? There is in fact a popular notion of something scholarly about wearing eyeglasses. It seems thus in reality not far-fetched to explore the possibility that myopia may have attained high rates in more developed countries because it carries with it some intellectual advantage.

A second likely candidate is the gene thought to be involved in the tendency to alcohol abuse. The problem of alcoholism certainly is very common, and familial trends seem well established. The opinion is widespread that social leaders and successful businessmen often have drinking problems. Perhaps they occupy their positions in society as a benefit of some form of brain stimulation, caused by an alcohol gene which, however, has a price attached.

Additional frequent genes that conceivably could have an intelligence effect are those involved in diabetes, epilepsy, or coronary heart disease.

9
Evidence for a Myopia Gene

Everyone knows that some individuals become nearsighted, being unable to see clearly objects at a distance, while close vision is unimpaired. Such individuals must wear eyeglasses, usually starting in adolescence.

Nearsightedness has for some time been known to run in families, but many eye specialists have preferred the explanation that excessive close use of the eyes is the common factor, the disorder hence being caused by strain from near work. Several investigators have attempted to document this hypothesis, but no one has successfully demonstrated a true relationship of eyestrain to visual impairment. Other proposed factors are poor lighting, improper printing, or forward tilt of the head during reading.

Early reports describe an excess of myopia in craftsmen, such as watchmakers or diamond cutters, but the question always remains whether the work results in myopia or whether those with a myopic tendency choose that kind of occupation. Research on school populations has documented that students tend to manifest a progressive increase in myopia, which can continue past age 20 years. Then there are recent reports on Eskimo and Indian populations in Alaska and northern Canada, demonstrating an "epidemic" of myopia among the younger generation, while the older members are almost free of the disorder. These reports have helped to revive the interest in environmental causes of human myopia.

Government-supported research is at present in progress, attempting to demonstrate that monkeys develop nearsightedness if they are reared under conditions preventing all distant vision. In reality it seems dubious that such a model reflects a mechanism paralleling human experiences. Many gene-mediated developmental processes are dependent on interaction with the environment, so that it is not surprising that disturbances can be produced by active interference with the normal sequences.

Because of the persistent interest in environmental theories, it is necessary to review in detail the evidence for the existence of a myopia gene.

Population differences in myopia

It has been known for some time that myopia tends to be concentrated in the more developed areas of the world. The highest rates are reported for groups of Jewish, Japanese, and Chinese origin, in the order of 25 percent, followed by western Europe and the United States, with only slightly lower rates. Systematic studies of randomly selected groups in underdeveloped countries are scanty, but some areas, for example in central Africa, show rates well below one percent.

The term "school myopia" has been widely used to refer to the increased rate in populations engaged in close work. Obviously there are many persons who read extensively without becoming myopic, so that near work cannot be the only factor. One study has demonstrated that illiterate Chinese women developed just as much myopia as the males who practiced reading.

As one would predict, myopia is infrequent in populations that depend for their livelihood on hunting or food gathering. It is not clear at what time the transition took place toward higher rates of nearsightedness in the more industrialized nations.

Myopia is rare before age eight, but increases particularly during the period of puberty. This indeed coincides with the school period, the condition seeming therefore to develop in parallel with increased demands on the eyes.

Myopia in twins

An obvious approach to an assessment of genetic factors is to

Evidence for a Myopia Gene

compare twins for development of myopia. Studies of foster-reared individuals do not seem called for, as no one has seriously proposed that the disorder is transmitted through environmental contact within families. The suggestion was, however, made by one of the groups studying Eskimos that some mothers were more likely than others to assemble their children around a study area, which was known to be poorly lit, and this was supposed to account for the observed concentration of myopia in certain sibships. Such a proposal can hardly be taken seriously.

There are few twin studies available which were specifically directed at myopia. Eye specialists tend to think in terms of ocular components and are therefore likely to tailor their twin studies toward measurements of similarities of the various specific structures, rather than dealing with the broad subject of myopia. As a consequence, some of the twin data do not even identify the myopic twins, but instead tabulate measures such as diopter differences in lens refraction or corneal curvatures for all monozygotic twins, mixing together the various eye conditions. When the data on the myopic pairs are specifically given, often one must select out the few myopic pairs from a large study of mixed data on twins.

A survey of the literature in connection with the present study was confined to twin pairs in which the index case had a myopia documented to be at least one diopter bilaterally. The study was divided into twins from individual case reports, of which there were 21 monozygotic and two dizygotic sets, and those encountered in surveys of twin groups, totaling 78 monozygotic and 37 dizygotic pairs with myopia. Of the 99 monozygotic pairs thus located from the literature, all but six were concordant for the disorder. Most of the discordant pairs showed mild myopia in the index twin, and only one monozygotic pair with more severe myopia appears to be definitely documented to have been truly discordant.

These data were supplemented by a small new study done at the Napa High School in California, identifying seven monozygotic and two dizygotic pairs. The monozygotic pairs were all concordant, the dizygotic pairs discordant. This brings the total data to 106 monozygotic pairs, 100 of them concordant, and 41 dizygotic pairs, 12 concordant.

In view of the very high concordance rate in the monozygotic twins, while the dizygotic pairs show no higher rates than those usually observed in sibs, it seems far-fetched to propose that environmental factors can be the chief causes of myopia. Even if one accepts the argument that monozygotic twins spend more time together than dizygotic pairs, it is difficult to envisage environmental causes that would almost invariably afflict both or neither monozygotic partners. The twin data therefore give very strong support to genetic factors as the principal cause of myopia.

Genetic mechanism for myopia

Accepting that true myopia indeed appears to be of genetic origin, one needs to know the mode of transmission to be able to define properly the type of gene involved. Among those eye specialists who favor a genetic cause for nearsightedness, there still is no agreement about the exact mechanism. Textbooks on eye disorders usually state that the transmission varies in different families, being dominant in some, recessive in others, and sometimes even sex-linked.

A geneticist is likely to view these claims with skepticism. If the disorder is in fact inherited and as frequent as the reports indicate, one must postulate that the usual factors involved in genetic polymorphisms are operative. It seems quite unlikely that there exists more than one pleiotropic gene, having some unknown favorable effect, as well as influencing the eye in a myopic direction. The proposal is obviously much more acceptable that just one such gene occurs at a high frequency.

Eye specialists also tend to arbitrarily divide myopia into the mild type below two diopters or 2 D, moderate from 2 to 6 D, and severe above 6 D. Various textbooks state that in general the severe type is inherited as a recessive condition, while the other types may have more than one cause. Although the above division is convenient, there actually exists no biologic basis for such a separation.

Review of the literature on genetic mechanisms for myopia leads to the conclusion that there is essentially no support for sex-linked inheritance. The evidence for dominant transmission in some families is likewise very unimpressive. Some authors have simply classified as dominant all families showing uninterrupted myopia in two or more generations. In reality such an occurrence is also consistent with recessive transmission, when one is dealing with a frequent gene, this having been referred to as pseudodominance. No systematic study has actually yielded data supporting dominant inheritance, and such a mechanism has only been proposed for some families, all investigators agreeing that recessive transmission indeed occurs in many kinships.

When one evaluates the more quantitative investigations, these are found to fall into two groups. In one type, data have been gathered on large populations, analyzing whether the findings are consistent with the expectations based on some standard genetic system. The best study of this type was reported by Furusho in Japan, but unfortunately the data were based mainly on school-connected families, the children therefore being relatively young. Still, only individuals above age sixteen years were actually included in the final study, embracing sibs and parents. Deriving formulae for expected rates in various groups on the basis of different hypotheses, Furusho demonstrated that his data were consistent with recessive transmission. However, he had to propose only 60 percent penetrance to account for the findings, and it should be mentioned that he excluded from the myopic group anyone with myopia of less than 2 D. In reality this approach is an entirely sound research strategy, and the stipulations result in a great reduction in the complexity of the calculations, but one must be aware that incomplete penetrance can be an artifact of such a design. For example, a sixteen-year-old individual with a myopia of only one diopter is likely to eventually become a true myope, but he would be considered a nonmyope in this study. The findings indeed give definitive support to recessive transmission as the major mechanism, but the question of incomplete penetrance is in reality left unsettled.

Another similar study had previously been performed by Paul in Germany, but his investigation was of less precise design. All persons in a large population were surveyed, without indicating which were young children or what degree of myopia was seen. When analyzed with Furusho's method, the data, however, fit well a recessive hypothesis, the findings in reality being surprisingly similar in the two investigations.

If one provisionally accepts the recessive hypothesis as a working model, an alternate test of its validity is to study the offspring of different types of matings, and some authors have approached the problem in this manner. Since families at risk can generally only be identified when a myopic child appears, the logical design is to attempt to determine the rate in the sibs of index cases, dividing the families into those with both parents unaffected, those with one parent affected, and those in which both parents are myopes. The theoretical risks in these three groups are 25, 50, or 100 percent if penetrance is complete, still maintaining the 1:2:4 ratio even if expression should be incomplete. Since some families at risk are always lost, namely those which by chance produce no affected child, the computations must take the resultant distortions into account. In connection with the present research a limited survey was made according to this design, covering only adult family members. The rates in the three groups were found to be close to the overall expectations, indicating actually complete penetrance. A larger similar study was reported in Germany many years ago, again consistent with a risk of one-fourth in sibs with unaffected parents and one-half when one parent is myopic.

Unfortunately, eye specialists do not routinely collect data on whole families, only seeing those members who have visual problems. Their data would also tend to be distorted by the type of sampling involved, which follows no standard design. Further studies of the above type are needed, but to be of real value they must be done correctly. Limited family pedigrees are of little use, but an example is shown in Figure 9-1 to illustrate how typically there is a skipping of generations in families with myopia.

The final convincing argument in favor of recessive inheritance derives from studies of children born to two myopes. However, since the distribution of visual measures in the normal population follows a bell-shaped curve, centering around zero diopter correction and extending on either side to approximately two diopters, it is necessary to limit this type of study to parents with bilateral nearsightedness of at least two diopters to ensure that they are true myopes. But since the myopic population overlaps with the normal one, extending from zero diopters upwards, with a peak at 3 to 4 D, all children of two definite myopes must be counted as myopic if any significant degree of refractive error is found. Furthermore, since in most cases myopia becomes manifest after age eight to ten years, its development often being incomplete even at age twenty years, only adult offspring should be counted. Following these rules, a sample of over fifty offspring of two definite myopes has been studied in connection with the present research, and all have been found to be myopic. If these findings are indeed representative, the data are only explainable in terms of recessive inheritance, as no other form of family transmission leads to a one hundred percent rate of the abnormality in the offspring. Although older data on matings of two myopes, which did not follow the above design, had shown somewhat lower rates, the evidence for a myopia gene, which leads to recessive transmission of the eye defect, can now be considered substantial.

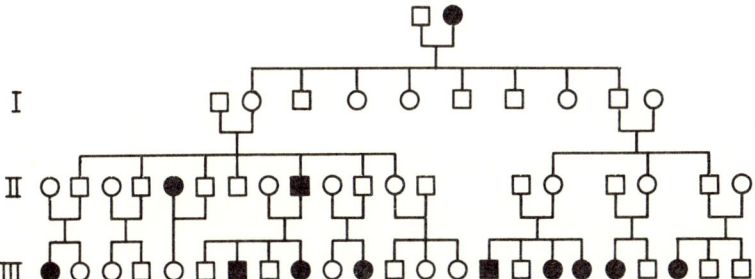

Figure 9-1. Pedigree of a family descended from a myopic index case, squares indicating males, circles females. Starting at top, rows of individuals represent ancestral couple, their children, grandchildren, and great-grandchildren, mates of those married also included. Myopic individuals are shown in black.

Myopia is one of the few examples of a metric trait associated with a recessively transmitted gene. To some extent, however, this is an illusion, produced by other overlapping effects. Many factors influence the total refractive power of the eye, leading to a bell-shaped distribution for each genetic population. The emmetropic group thus shows a range from +2 to -2 D, these apparently being the limits beyond which aberrations of such structures as the cornea or lens are unlikely to go. The true myopic group ranges from zero diopters upwards, increased axial length caused by the myopia gene in a homozygous state usually carrying the correction past 2 D. In populations with many myopes, the overall distribution curve is consequently seen to be bimodal, although the distribution is still uninterrupted. In a population with a low rate of myopia, visual measures show only one continuous curve, the relatively few true myopes then being blended in with others at one end of the curve.

Opposition to the genetic view

It should be reiterated, despite the seemingly convincing genetic evidence, that there still is considerable controversy even about the hereditary nature of myopia, not to mention the transmission mechanism. For example, the recent reports on Eskimo populations, revealing a high rate of nearsightedness only in the younger generation, have been interpreted by some investigators as support for an environmental etiology. For a geneticist it is, however, tempting to explain these data in terms of natural selection. If myopia was in the past essentially incompatible with survival under the conditions that these people had to adapt themselves to, few adult Eskimos should be myopic. To account for the high rate in the new generation, one must then postulate that under extreme conditions natural selection favors survival of the carrier of one myopia gene.

The studies of Eskimos have shown an increased myopia risk in the sibs of myopes, an observation consistent with recessive inheritance. Although the investigators reporting these findings have used expressions like "environmentally induced epidemic," it seems improbable that myopia has a different basis in Eskimos than in other human populations. The total evidence in favor of heredity is so convincing that the somewhat puzzling aspects of the Eskimo data presumably must be accounted for by differential selection. The new findings can then in reality be interpreted as support for the view that before the advent of eyeglasses the myopia factor indeed fell into the class of polymorphic genes with a favorable overall effect only in the heterozygous carriers. During that period nearsightedness was presumably a debilitating associated condition, occuring in homozygous individuals who had limited potential for survival.

Criticism of those who see evidence for heredity of myopia seems to originate even from lay groups, which claim to be promoting prevention of eye defects. Without attempting to evaluate the scientific evidence, these groups support the view that heredity plays a very limited role, also expressing the hope that public funds will not be employed to explore the genetic theory. Their leaders can rest assured that no financial support was used in the present study, and there were no ulterior motives, just scholarly interest. The main officially supported approach in this field is indeed the environmental one, and for some reason that effort has been led by a psychologist.

Despite opposition by various critics, the scientific evidence seems persuasive, giving substantial support to genetic factors and specifically indicating recessive inheritance, probably with full penetrance.

10
Myopia and Intelligence

Having established, hopefully to the satisfaction of most readers, that nearsightedness indeed has a genetic etiology, we can now return to the question whether the myopia gene is involved in intellectual development. As was indicated earlier, a hereditary condition existing at a high frequency is likely to involve genetic polymorphism, making the question of an association with intelligence a fully legitimate one.

Increase of myopia in groups of higher performance

Early reports of a concentration of myopia among intellectual groups date at least as far back as 1813, when it was observed in personnel of the British navy that the officers were more often nearsighted than the men of lower rank. Half a century later the medical literature describes a progressive increase of myopia in the brighter pupils in northern European schools. It has thus been known for well over a century that myopes outperform nonmyopes in academic types of achievement. This finding has been so consistent and so definitive that all published reports are in substantial agreement. In some of the studies, total school populations have been examined and separated into myopes and nonmyopes. School grades were then compared for the two groups. In other investigations the top performers were compared with the less successful students in terms of myopia rates. In Western countries it turns out that a large number of students who are admitted to universities are myopic. In Iceland almost one-half of the honor graduates from the college preparatory schools wear glasses for myopia. At the University of California at Berkeley over half of the students are nearsighted, indicating that myopes in that area are five times as likely to succeed in the entrance examinations as nonmyopes, since they are outnumbered by that factor in the overall population. The Terman study of gifted children in California similarly revealed a high rate of myopia.

More recent reports deal in IQ scores. Hirsch published a paper in 1959 on groups of American shool children who had been given complete eye examinations. Intelligence test data were also available, showing that the nearsighted group achieved the highest scores, followed by students with normal vision, the farsighted group in this case falling below the other two. However, the samples were relatively small. This paper was followed in 1970 by a report from New Zealand, also dealing with school children, but again based on a rather small sample. The results of that study, reported by Grosvenor, utilizing the Raven Matrices and Otis psychologic tests, yielded comparable data, suggesting the same order of performance for myopes, emmetropes, and hypermetropes.

Since these studies were generally carried out by eye specialists, they naturally concentrated more on the eye data than on other aspects of the investigation. However, in the case of school children it is relatively unimportant to know the precise degree of refractive error, since this is likely to change rapidly anyway in that age group. Separation of definite myopes from the remainder of the population is all that matters at this stage for a geneticist interested in the reality of the IQ difference. The disadvantage of insisting on exact eye data is that the investigators using that approach have had to settle for relatively small groups of students.

To take advantage of a larger group, the myopia phase of the present investigation was directed at a complete survey of a population of California students in a community of approximately forty thousand inhabitants. The study embraced all pupils aged seventeen to eighteen years, covering a three year period. This group included most persons of that age in the community, as school attendance is required by law until age seventeen years, and special classes are provided for mentally retarded individuals. Almost all the students in this community are white, including only a few of Mexican-American descent.

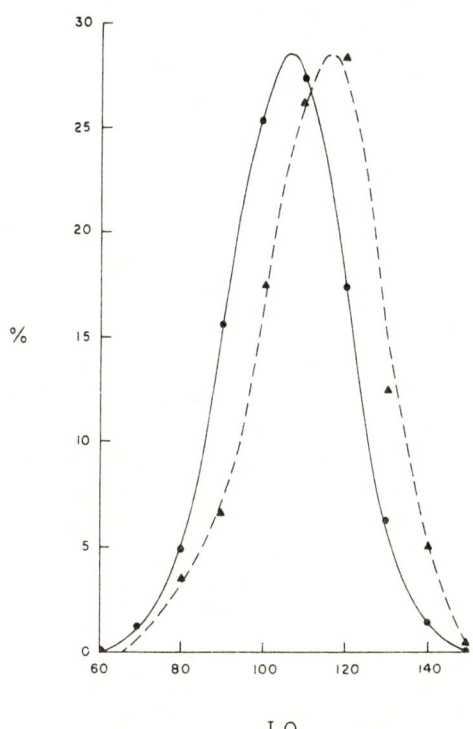

Figure 10-1. IQ distribution in a school population at age seventeen to eighteen years, showing separately the overall scores (solid line) and the values attained by those individuals who were myopic (broken line).

The student health service had information on all the students, as periodic checks are made to ensure that eye problems are not interfering with school progress. These data would no doubt be considered incomplete by eye specialists, but in most cases they were sufficient to decide whether or not a visual problem existed and whether the problem was myopia. When necessary, it was possible to contact the student or his family for additional information.

In this manner the 2527 students gathered over the three year period were found to include 377 myopes, a rate of 15 percent, which is consistent with the reported rate of the disorder for the area under study. It seems probable that essentially all the students classified as myopes require a correction of at least one diopter, but the precise degree of myopia is unknown.

Since at the time of this study all California high school students in their senior year were required to take a group IQ examination, the scores on the Lorge-Thorndike test could be compared for the myopes and nonmyopes at age seventeen to eighteen years. The distribution curves were found to be identical in shape, but the curve for myopes was shifted eight points toward the higher range, as is illustrated in Figure 10-1. Furthermore, since it is likely that most myopes in this school population have indeed been identified, it is possible to compute the rate of myopia in the different IQ groups. This information is shown in Table 10-1. It is of interest that the group of the so-called cultural familial type of mental deficiency, including only thirty members in this population, is devoid of myopes. This is in reality to be expected if the IQ curve for myopes is shifted over eight points. Other studies have shown that severe forms of mental deficiency, specifically mongolism, include just as many myopes as the general population.

Unfortunately the data do not give an answer to the question whether higher degrees of myopia are associated with a still greater gain in intelligence. Judging from the number of high myopes seen on university campuses, one may conjecture that such a relationship may indeed exist, but scientific data are at this time unavailable, except for a suggestion of this finding in the study reported by Hirsch.

In view of the consistency of the findings of all investigators, there can no longer be any disagreement about the reality of the association of myopia with improved academic performance and elevated IQ scores. Interpretations of these findings are, however, far from being uniform.

Genetic interpretations of the correlation of myopia with high IQ

The commonly preferred explanation of the increased rates of myopia in higher IQ groups has been that a myopic student, because of his physical limitation, tends to stay indoors and study, thus sharpening his scholastic acumen. Fortunately, this type of hypothesis can be tested with available data. The information contained in the cumulative records of the California students included IQ tests administered at younger ages. Examination of this material revealed that those students who had become myopic at 17-18 years had already achieved the full IQ advantage by age eight years, that is before nearsightedness generally develops. Precise information has been provided by Hirsch on California school children, showing that myopia exceeding one diopter exists by age eight years in less than one percent of the population. The IQ gain thus precedes the development of nearsightedness.

The genetic studies reviewed in the preceding chapter leave little doubt that myopia is dependent on hereditary factors. How else does one explain the essentially complete concordance for myopia in monozygotic twins or the adherence of the family distribution to a simple algebraic system, recognized as a standard genetic mechanism?

In reality it is immaterial, for the purposes of explaining the link between myopia and intelligence, whether the mode of transmission is accepted to be recessive. If myopia is in some manner related to a specific gene, regardless of the exact mechanism, and the carriers of this gene show an intellectual gain as the first manifestation of its presence, the obvious explanation is that the gene somehow affects brain function.

Table 10-1

Rates of myopia among different IQ groups in a northern California school population aged seventeen to eighteen years

IQ range	Number	Myopes %
55-74	30	0.0
75-94	516	7.4
95-114	1317	12.5
115-134	619	24.9
135-	45	44.4

It is easy to visualize how the myopia gene may actually be principally a brain gene, exerting its effect primarily on the central nervous system. The resultant stimulation then leads to enhancement in school performance and upward shift in IQ measures.

Accepting that this has been established to occur in the person homozygous for the proposed gene, one can legitimately wonder if perhaps there is also some stimulant effect in the heterozygous state. If there should be a partial effect also in the carrier, this would help to explain the high rate of myopia in young Eskimos, as was mentioned earlier. Eskimos have struggled to survive in regions that most humans would consider uninhabitable, and their margin for survival has been so narrow that they even had to develop systems of actively disposing of unfit persons to avoid any unnecessary burden. Under these circumstances it is not unlikely that only the fittest could survive, and that may have included the myopia carriers. Other northern countries, where life has been highly competitive and no doubt difficult for persons with a visual handicap, also have a surprising frequency of the myopia gene. Thus Iceland shows a 30 percent gene frequency or 9 percent rate of myopia. It is hard to imagine that myopes enjoyed a survival advantage in Iceland, and one wonders whether the myopia carriers didn't show increased fitness.

Direct measures of the IQ of myopia carriers are unfortunately not available. This information could be obtained, for example, by a study in a given area of all nonmyopic persons who have one myopic parent. Investigations of that type hopefully will be done in the near future. In any case, the carriers must have a lower IQ than the myopes, or else the latter could not become as concentrated as they are found to be in academic groups.

The increased frequency of myopia in industrialized societies also supports the genetic nature of the intelligence association. No advanced industrialized culture exists without the population showing an elevated incidence of myopia. The entire constellation of findings can hardly be just fortuitous. The logical conclusion is that the myopia gene has an important stimulant effect on brain activity. It thus becomes the first identified specific gene which appears to contribute significantly to intelligence.

11
Existence of an Alcoholism Gene

Although the indications seem quite convincing that the gene involved in myopia exerts a stimulant effect on the brain, it is evident that only a fraction of superior mental ability is accounted for by this factor. Many successful individuals, such as those who assume leading positions in government, industry, commerce, professional activities, or even in academic life, are nonmyopes. Obviously they have other things going for them.

It is of some interest that there appears to be a common bond among many persons in the above groups in the sense of being self-confident and relatively aggressive. It is also not uncommon for socially successful persons to develop a taste for alcoholic beverages. The recognized problem with alcoholism among business executives, as well as among politicians and various professionals, may reflect a personality makeup shared by the leadership types of individuals. Since there are definite indications that alcoholic tendency runs in families, it seems worth while to investigate more specifically what factors may be operative.

Support for inheritance of alcoholic tendency

It has been estimated that in the United States three-fourths of all alcohol consumed is imbibed by one-sixth of the people, and certainly some families are more heavily involved than others. Since either genetic or environmental factors can conceivably account for a family concentration of a trait like alcohol abuse, the significant types of study must include data on twins and foster-reared relatives. Both kinds of information have been supplied.

A study reported from Sweden compared 174 twin sets in concordance for alcoholism. In each case the index twin had a well documented alcoholic history. The 126 dizygotic pairs were found to have an alcoholism concordance of 28 percent, while the 48 monozygotic pairs exhibited almost twice that rate or 54 percent. Even among the discordant sets, the degree of difference was considerably less for the monozygotic twins. Another twin study conducted in Finland, involving 557 dizygotic and 172 monozygotic pairs, led to heritability estimates in the order of 50 percent for amount of alcohol consumed and frequency of drinking episodes. This latter study compared randomly chosen twins, rather than those with documented alcoholic index cases. Both investigations give definite support to hereditary factors in alcohol abuse.

In contrast to these carefully conducted and well controlled studies based on twins, an evaluation of the literature on foster reared children of alcoholic parentage illustrates the pseudoscientific approach commonly utilized by those inclined to promote environmental viewpoints. For many years various reviewers of the evidence for hereditary factors in alcoholism have quoted the one supposedly significant study of this type, conducted in the United States and said to have disproved the likelihood of any genetic influence. This study, reported by Roe and Burks over thirty years ago, involved a comparison of thirty-six children of alcoholic parents and twenty-five control children, both groups raised in foster homes. At average age of thirty-two years there was stated to be no demonstrable difference in alcoholism.

Examination of the original data reveals that just twenty-seven and twenty-two members, respectively, of the two reported groups were in fact included in the alcoholism study. Furthermore, many of the subjects were girls, and there is no point in even talking about rates of alcoholism in a few relatively young women, since these groups are far too small for any meaningful study of female alcoholism. Only twenty-one versus eleven cases in the two original groups were males, but unfortunately, it is not stated how many of them ended up in the final alcoholism data. The number of actually studied foster reared sons of alcoholics is probably well under twenty, and none are said to be alcoholic as adults, although at least two had earlier been in difficulty with alcohol abuse. The expected number of true alcoholics at age thirty-two in such a sample, to be compared to less than 11 controls according to a genetic hypothesis, is however, so low that nothing of significance can be established. Furthermore, the purported alcoholism of the fathers was not well documented. This study should not have been used for a third of a century as an argument against hereditary factors, but it seems to have served the needs of environmentalists in charge of research funds, maintaining the status quo and discouraging further genetic investigations.

A new and highly significant investigation of foster reared children born to alcoholic fathers has recently been conducted in Denmark. Apparently these fathers now have a well documented history of alcoholism, hospital treated, although their illegitimate sons, given up for adoption many years ago, were presumably born mostly prior to that diagnosis. The fifty-five sons in the experimental group were compared to seventy-eight males born of nonalcoholic parents, who were placed in matched foster homes. Examination for alcoholism was performed on both groups by psychiatrists who had no knowledge of which type of parents the individual belonged to, and the information could not have been obtained from the subject himself, as he didn't have that knowledge either. A highly significant difference in alcoholism was revealed when the data were finally analyzed, showing 18 percent and 5 percent rates in the experimental versus control groups. The differences in alcoholism requiring treatment were even greater.

Another important study has been published recently in the United States, comparing the rate of alcoholism in half-sibs arising from parents who had had more than one marriage partner. In essence the study separated four groups, those with a biologic parent who was alcoholic, having been reared with or away from that parent, and those with neither biologic parent alcoholic, but reared either with an alcoholic stepparent or without exposure to alcoholism in the home. The results are quite clear, showing high rates of alcoholism in the sons of alcoholics, whether or not that parent was in the home, while no increase in alcoholism existed in sons of nondrinkers, even when they were reared by an alcoholic stepparent. The numbers in the different groups are sufficiently large to be highly significant. Almost one-half of the forty-six subjects born of an alcoholic parent were themselves alcoholic, including eleven of the twenty-two who were reared in the absence of an alcoholic parent figure.

Although the older studies, because of a less satisfactory design and a small number of male subjects, had failed to identify genetic influences, it is hard to see anything wrong with these more recent reports, which deal with sufficiently large groups. It seems justifiable to conclude that the operation of genetic factors has now been established.

Family distribution of alcoholism

Geneticists have recognized for some time that an increased risk of alcoholism indeed exists in the relatives of alcoholic patients, but few satisfactory quantitative studies are on record. One of the best investigations of the family distribution is that done by Amark in Sweden, which showed a risk of 26 percent in fathers and 21 percent in brothers. If this is compared to the 3.4 percent rate estimated for the overall male population by Fremming in another Scandinavian study, the rates in fathers and brothers are increased 7.6-fold and 6.2-fold, respectively. Similar increases were seen in mothers and sisters, as compared to the overall female population, although the total rates were less than a tenth of the male rate and therefore not ascertainable with the same degree of precision. These findings are in general accord with the results of many other authors who have gathered family data on alcoholism.

A preliminary family study of alcohol addiction in the male population of Iceland has been performed in connection with the present research. The design of this investigation started with identification of male index cases admitted to the mental hospital in Reykjavik and assigned a diagnosis of alcohol addiction, this being the only mental hospital in Iceland. The general population was divided into two generations, the first born between 1881 and 1910 and the second in the interval 1911 to 1940. The likelihood of ever having been admitted to the mental hospital for alcohol addiction was, respectively, 0.9 percent and 1.5 percent for these two age groups.

Existence of an Alcoholism Gene

When male relatives had also been identified, their rates of alcohol addiction were similarly ascertained, again classifying as alcoholic any person who had ever been admitted to the same hospital for a drinking problem. The results are expressed in multiples of the hospital admission rate for the general male population born in the same period and are shown under the heading "comparative risk" in Table 11-1. The rates are averaged for the two generations in the case of those groups which are represented in both. The figures are actually similar as regards the risk in fathers and brothers to the comparative rates estimated from Amark's study.

Using data previously published by Helgason on total population rates of alcohol abuse in Iceland for males born in the period 1895-1897, the comparative rates can be converted to absolute values, as experienced for those born in that interval. The results on true addiction are shown under the heading "alcohol addiction risk" in the table. In the last column all the comparative risks are multiplied by the factor 6.5, reported by Helgason as the rate of combined alcohol addiction and habitual excessive drinking for the Icelandic male population.

Table 11-1

Estimated rates of alcoholism in male relatives of alcoholic index cases in Iceland

Relationship	Number studied	Comparative risk	Alcohol addiction risk (%)	Total alcoholism risk (%)
Fathers	86	5.2	14.0	33.8
Brothers	320	5.7	15.4	37.0
Sons	53	5.8	15.7	37.7
Uncles	232	1.4	3.8	9.1
Nephews	122	2.2	5.9	14.3
First cousins	748	1.6	4.3	10.4
General male population	10,786	1.0	2.7	6.5

One notes that the alcoholism risk is reduced with a decreasing order of genetic relatedness, being highest and relatively uniform in first-degree relatives (fathers, brothers, sons) with proportionate decrease in second and third-degree relatives. It should be noted, however, that the total number of relatives in each group is rather small, permitting a considerable degree of statistical scatter and leaving the final figures subject to change by further research. Since availability of alcohol in Iceland has been restricted in the past, it is likely that the true addiction rates would be higher for the population born during the present century.

The results from Iceland are in satisfactory agreement with other data on the family distribution of alcoholism. The overall impression is that the true risk of alcohol abuse in first-degree relatives exceeds 30 percent in populations with liberal access to alcohol, and rates in second- and third-degree relatives are also rather high. However, several studies, including the present one, agree that risks of psychosis are not increased in relatives of alcoholic index cases, suggesting that alcoholic tendency is independent of other forms of mental illness.

Genetic mechanism for alcohol addiction

On the basis of the limited family data, it is not possible to arrive at definite conclusions about a genetic mechanism for the tendency for alcohol abuse. The family figures are, however, suggestive of dominant inheritance with incomplete penetrance, although they can also be explained by polygenic transmission.

Genealogic information available in Iceland makes it possible to separately evaluate a polygenic hypothesis. In a disorder existing at a very high frequency, a proposed scheme for polygenic inheritance does not involve disease-related genes, rather the postulation is that an unfortunate combination of otherwise innocuous genes is responsible. In the case under discussion this presumably results in a state of increased tension, which in turn causes an individual to seek relief in drinking. In view of the high general rate of alcoholism, all genes involved in such a system must, however, be very frequent, existing everywhere, rather than being unevenly distributed in the population. It is then merely a matter of chance when the unfortunate combination happens to show up. Consequently, there should be no significant difference in alcoholism rates between different kindreds or between different branches of the same kinship, although once the unfavorable combination occurs, it has an increased risk of recurrence in the close relatives.

To test these predictions, two well-studied kindreds, which

arose from parents living in Iceland shortly before 1800, were investigated. Both are from the same area, and the complete listing of descendants has been published in one genealogy book. The four branches derived from the children of one set of original parents contain between 141 and 292 males born in the interval 1911-1940, while the five branches in the other kindred show corresponding figures with 24 to 294 members. For the most part these individuals are removed four to five generations from the original parents.

The rates of alcoholic admissions to the mental hospital for the nine branches range from 0.0 to 6.7 percent, the lowest rate occurring in a branch with 292 members. A distribution as uneven as that observed is unlikely to be a chance occurrence in either of these kindreds, according to statistical tests of probability. This evaluation thus indicates that the polygenic hypothesis does not seem capable of accounting for the findings. Significant differences in alcohol addiction have also been observed between different kindreds in Iceland.

At this point, consequently, the best hypothesis appears to be one based on a single major gene, following a modified dominant type of transmission. Because of the far greater problem with alcoholism in males than in females, the possibility of sex-linked transmission has been explored, but all such studies lead to the conclusion that the alcoholism gene is not located on an X chromosome. The sex difference must have some other explanation.

12
Brain Stimulation Associated with Alcoholism

Since the new evidence supports the existence of a gene that predisposes its carriers to alcohol abuse and the frequency of this gene must be postulated to be quite high, some beneficial effect has to be associated. A search is thus in order for a possible relation of this entity with behavioral traits. Obviously there are various factors involved in personal decisions about abstinence, social drinking, habitual drinking, or compulsive drinking, and true pathology may exist only at the extremes.

The Finnish twin study (see p. 80) suggested that abstinence may have a hereditary component, and this presumably involves factors entirely separate from addiction tendency. The concern here is only with an alcohol addiction gene, postulated on the basis of the data discussed in the previous chapter.

Relation of childhood hyperactivity to adult alcoholism

A link seems to have been established between hyperactive behavior in childhood and alcoholism in the adult. Long-term follow-up studies of children exhibiting the type of restlessness now often referred to as "minimal brain dysfunction" have suggested that they often develop into adults with sociopathic or drinking problems.

Some investigations in this field have been directed toward

hyperactive behavior as such, without attempting to tie it in with alcoholism. Thus it has been demonstrated that relatives of hyperactive children show an increased rate of hyperactive behavior. More specific studies have involved twins and foster-reared children. Both these approaches seem to confirm that genetic factors indeed are operative, the literature having been reviewed recently by Cantwell. Heritability estimates for hyperactivity are reported to be in the order of 70 percent.

While Cantwell has proposed a polygenic model as the most likely mechanism for childhood hyperactivity, the data fit a hypothesis of dominance with incomplete penetrance just as well. With either mechanism the risk of hyperactivity in relatives of index cases should be approximately proportional to the degree of genetic relatedness.

A longitudinal study of hyperactive children was published several years ago under the title *Deviant Children Grown Up.* This investigation started out with children who had been referred to a clinic for behavior problems, where they were subjected to very complete diagnostic studies, excellent records being gathered also on the family composition and location of various relatives. The main thrust of the subsequent follow-up was to establish whether the problem persisted into adulthood, whether the usual therapeutic procedures had indeed made a difference, or whether the hyperactive behavior foreshadowed the development of mental illness. As is usually the case with hyperactive behavior, most of the children referred to the clinic were males. Despite the problems ordinarily encountered in the United States with a long-term follow-up, information was obtained on 90 percent of the original group after a lapse of thirty years.

The significant findings in this study relate to the type of behavior shown in adulthood by these individuals, selected in childhood on the basis of hyperactive or antisocial behavior. The important signs were a frequent occurrence of alcohol abuse and sociopathic tendency. There was no indication that any form of treatment had had any influence on the outcome. The frequent development of alcoholism suggests that brain stimulation in childhood, manifesting itself as hyperactive behavior, may be one of the signs of the presence of the postulated alcoholism gene. Although overstimulation of the brain is in this context seen as undesirable, it is not hard to imagine that the same effect might under other circumstances be viewed as beneficial. Hyperactive children, however, rarely developed into psychotic adults.

Adult hyperactivity and alcoholism

The term hyperactivity is usually applied only to childhood behavior, but there are indications that the same basic problem is to some extent responsible for antisocial or sociopathic activity among young adults. Crime is largely a problem of young males, and while mental attitudes obviously play a role, restlessness and boredom are no doubt important contributing elements.

Studies were reported many years ago on criminality in twins, the findings being that monozygotic pairs were frequently concordant for antisocial behavior. A book by Lange, entitled *Crime as Destiny,* described several such twins. The inclination has been to reject these reports as not representative, claiming that concordant pairs were specifically selected for inclusion in the studies. However, no actual data have refuted the validity of the findings, and subsequent reports seem to corroborate the conclusions.

A more recent study dealt with foster-reared children born of mothers who were incarcerated for crimes. Despite minimal contact with the mothers, the sons showed a significant increase in antisocial behavior. The available data, although obviously insufficient, thus support a genetic contribution to criminal tendencies.

The association of criminality with alcoholism is debated, but numerous studies suggest a relationship between the two conditions. As was mentioned earlier, the long-term follow-up of hyperactive children showed an increase in both alcoholism and antisocial behavior in adulthood.

Studies of foster-reared relatives of alcoholics have also shown an increased rate of divorce in such groups, this being one of the statistically significant findings. Marital instability may thus be another sign of hyperactive behavior.

Relation of alcoholic tendency to achievement

Since there is an impression that alcoholism is elevated in frequency, not only in persons with an antisocial history, but also among professionals and social leaders, a search seems in order for positive effects possibly related genetically to alcoholism. Background information already available on the population of Iceland, including the family data presented earlier, has made it possible to survey various groups of relatives of alcoholics for signs of social success.

The actual study was conducted as follows: To start with, randomly selected samples of males, born in the intervals 1851-1880 and 1881-1910, were assembled, partly from the data already utilized in Table 11-1. Both age groups were studied for inclusion in a compendium of *Who's Who in Iceland,* published in 1944. The rates for the general population were thus found to be 12.0 percent and 7.0 percent, respectively, indicating that in a small country with a high index of culture such listings cover a large portion of the population. Various groups of male relatives of alcoholic index cases were then subjected to the same kind of study, with the results shown in Table 12-1. Alcoholic patients do not show an increased rate of listing, and this is understandable, since such a group includes many hard-core alcoholics, who are not likely to be considered eminent. However, all groups of relatives show elevated rates, and these are statistically significant. Since fathers are a special group, probably not entirely comparable to the general population, their elevated rates of listing may not be entirely unexpected. Nevertheless, it appears that the proposed alcohol gene may in fact lead to some form of brain stimulation, which increases the likelihood of inclusion in *Who's Who.* This finding is in harmony with the overall impression that there is an abundance of successful persons in families of alcoholic patients.

A separate study was done of the top graduates of the college preparatory school in Reykjavik to ascertain whether their first-degree relatives might show excessive rates of alcohol addiction. No such increase was found. Whatever it is that leads to elevated rates of inclusion in *Who's Who* of the relatives of alcoholic patients thus does not appear to involve distinguished scholastic performance.

In his Swedish investigation, Amark attempted in different ways to assess the mental characteristics of alcoholics. On certain intelligence tests, such as word associations, these individuals indeed showed above average performance, but the differences between groups were not sufficient to reach statistical significance. There is an impression, however, that verbal fluency may be characteristic of alcoholics. It is known that famous authors are likely to have problems with alcohol. Upton Sinclair wrote a book describing alcoholism among his renowned colleagues, and it has been found that Nobel laureates in literature often have had difficulties with excessive drinking.

Table 12-1

Comparison for males of rates of listing in *Who's Who in Iceland*, covering the general population, alcoholic patients, and relatives of alcoholic patients

Type of individual	Males born 1851-1880		Males born 1881-1910	
	Number	% listed	Number	% listed
General population	1480	12.0	3456	7.0
Alcoholic patients	—	—	213	6.6
Fathers	125	25.6	86	12.8
Brothers	—	—	160	12.5
Uncles	—	—	232	11.6
Grandfathers	146	13.0	—	—
First cousins	—	—	236	8.5

Much more research is needed before solid conclusions can be reached about an association of alcoholism with giftedness or leadership qualities, but the new data are indeed suggestive. It would be helpful if objective tests could be devised to quantitatively assess the personal qualities exhibited by families with alcoholism.

If one accepts that success in the business or leadership arena demonstrates a measure of giftedness, it seems that the proposed alcoholism gene indeed is related to some form of enhancement of brain function. Perhaps society has in the past unfairly condemned and degraded families with tendencies to alcoholism. Their genes may be important for the maintenance of the complex structure of an industrial state. Further studies are needed to establish the exact nature of the intellectual advantages which such families appear to harbor.

13
Physiology of Learning Functions

Chemical studies of learning mechanisms are in their infancy, but it seems probable that major advances in this field are just around the corner. All normal persons appear to possess the same basic brain machinery, so that ordinary variations in intellectual powers are not based on gross qualitative differences in anatomy or physiology. The degree of facility with which the machinery is available for use, perhaps the nature of the tuning mechanisms, so to speak, appears to be at the root of many differences.

Most investigators agree that important learning functions are likely to be located mainly in the cortical areas of the cerebral hemispheres, but there is no doubt that emotional aspects, under control of lower centers, are also involved, at least in facilitating learning. Experience which has an emotional content has an increased likelihood of being remembered.

Some of the theoretical models of the brain have been constructed through analogy with computers, storage and retrieval of information then being viewed as largely dependent on mechanistic types of function. In explorations based on such models, attempts have been made to separate different aspects of the learning process, for example the short-term impression versus the more permanent memory storage. However, such exploration has met only with limited success, and the value of the computer analogy is now questioned by many investigators. Newer evidence indicates that a widespread participation of various regions of the brain is demonstrable in connection with different learning functions. Specific types of learning show greater involvement of some areas, without the definitive kind of localization of different processes that once was believed to exist.

Specific chemicals in learning

Various suggestions have been made for definitive substances involved in storing and mobilizing information in the brain. One hypothesis has been that ribonucleic acid (RNA) may be a factor, but this has proven difficult to substantiate. Some early experiments with metazoa did suggest that certain flatworms could consume RNA from trained members of the same species and thus acquire new information. However, in the final analysis the subject becomes considerably more complex. One proposal is that preformed building blocks may be of benefit and on the surface this may look like acquired knowledge. Suggestions have also been made that RNA has a facilitating or a reinforcing function in learning.

Recent research in several centers has further supported the idea that RNA and protein synthesis may indeed be in some manner involved in the storage of information. Some of the studies have utilized isotopically labeled chemicals to demonstrate that increased synthesis of these substances occurs in the brains of animals, such as chicks or rats, exposed to certain learning experiences. Other investigators have utilized specific chemical inhibitors to show that learning can be interfered with by preventing synthesis of RNA or protein. Although there is some consistency in the results obtained by different investigators, it is difficult to be sure that the changes are indeed a part of the learning process itself and not secondary effects of the arousal and activity which are unavoidable when the animal is being trained.

Suggestions have been made that pharmacologic agents influencing RNA synthesis may be of benefit in older persons with memory loss. Some of the specific drugs recommended for this purpose have, however, later turned out to have other actions, being essentially central stimulants, rather than necessarily acting through RNA.

Recent animal experiments have revealed an enhancement of retention and memory by relatively simple polypeptide hormones, normally elaborated by the hypothalamus or the pituitary gland. Some of these are chemicals previously thought to increase arterial blood pressure (this raises the question of whether human hypertension may not be another condition associated with some beneficial mental traits). Others are chemicals which stimulate the cortex of the adrenal gland to secrete its hormones, the mental effects appearing to be separate from the control of internal secretion.

Chemicals in myopia and alcoholism

At present the knowledge about chemicals involved in myopia is limited. Several drugs are known to cause temporary myopic changes in the eyes of occasional patients, but this is in most instances thought to result from idiosyncracy of the particular individual in regard to the medication. Also, since permanent myopia is usually associated with measurable elongation of the eyeball, it is hard to see that lens changes would have a bearing on that problem.

Several reports indicate that local treatment with atropine slows down the myopic changes in the eyes of teenagers, and this has been confirmed by treating one eye, using the other as a control. Since atropine antagonizes the parasympathetic effects of acetylcholine, one of the important neurohormones, this raises the question whether the latter transmitter may be involved in myopic development. Thus one may conjecture that perhaps the myopia gene is associated in some manner with excessive production of cholinergic substances, but more precise studies are needed to test such a hypothesis. One can certainly hope that the identification of specific genes, which somehow enhance learning, will give new impetus to attempts to identify the responsible chemicals.

Many attempts have been made to search for chemical abnormalities associated with alcoholism, but at this time there is little evidence that observed alterations reveal the basic metabolic difference which presumably exists between normal and alcoholism-prone persons. One type of study appears to have established a link between alcoholism and the chemicals involved in color blindness, but the eye changes have been found to be a secondary effect, apparently caused by toxic effects of alcohol.

Most of the investigations have concentrated on the metabolic breakdown of alcohol itself, and no basic differences have been found between alcoholic and control populations in that regard. In reality there are no indications that the inherited metabolic aberration involves the oxidative pathways of alcohol in the body. Hopefully the demonstration of a definite genetic basis for alcoholism will help to lead investigators into new areas of biochemical exploration.

Some studies of alcohol addiction have been performed by the use of animal models. The obvious weakness of this approach is that there is no reason to believe that randomly chosen animals possess the same inherited abnormality that now appears to exist in human alcoholics. Whether "addiction" in experimental animals corresponds to the human condition remains an open question.

SECTION III
Biologic Factors in Creativity

14
Definition of Creativity

It is generally recognized that mankind's cultural progress has been dependent mainly on the contributions of a relatively small number of great men. While contemporary leaders often leave behind impressions of limited duration, only the highly creative make a permanent impact and alter the course of history. Men of creative genius arise at a rate of less than one in a million, but naturally there exist many with creative talents of lesser degree.

In the early days of intelligence tests, when there was still exuberant enthusiasm about their capacities, psychologists felt they had come upon a tool which, among its other potentials, could identify persons of the highest creative caliber. Those achieving an IQ score above 140 on the tests were during this period referred to as designated persons of genius. Extensive longitudinal studies of such individuals were commenced, with a plan to publish periodic reports on their progress through life. Funds were also allocated to provide special opportunities for students seen to have great promise on the basis of test performance.

Eventually, however, the sobering findings emerged that the individuals selected by psychologic tests from a large population, while unquestionably intelligent and socially competent, failed to evince the characteristics which are found in those truly creative men of genius generally credited with human progress. In fact, it almost seems that the test procedures, placing, as they do, a premium on robust health and competitive spirit, may have eliminated such persons of highly creative potentials as presumably existed in the large initial population. By now it is generally acknowledged that IQ tests do not measure creative capacities, and "giftedness" is seen as a more appropriate term to designate persons of high IQ.

Partly as a consequence of these manipulations with the concept of genius, the word "creativity" has now assumed greater importance in describing those qualities which enable some persons to visualize unrecognized associations and invent novel ideas. A definition with general validity is hard to coin, and even authorities on creativity tend to use the word in a way that happens to include their own personality traits. One kind of scientific approach has been to select individuals whom all consider creative, and through a study of their personality makeup arrive at data for a valid definition. Many tests of "creativity" have also been designed.

Studies of creative characteristics

The impression from newer endeavors in this field is that creativity indeed involves quite specific processes, but different from those required for superior school performance. Looseness in associative thinking, ability to tolerate ambiguities, and facility for making conceptual leaps are some of the descriptions encountered. Effective individuals are also characterized by solid ego strength, immense dedication to their work, and willingness to suffer deprivation. Many highly creative persons devote essentially all their free time to their chosen effort.

The creative process itself is often seen as an intuitive act, not necessarily logical, and only in part conscious. Often the solution of a problem is preceded by a period of subconscious incubation, the critical answer sometimes emerging suddenly, not uncommonly after a sleepless night. The critical point of a creative act frequently is accompanied by a sensation of inspiration or even ecstasy. Creativity thus has characteristics which bear little resemblance to the thought processes measured with customary intelligence tests. Wallas has described the creative process in terms of four stages, designating these as preparation, incubation, illumination and verification.

Creative aptitude, when combined with scholastic ability, generates the ingredients essential to intellectual genius. There exists an abundance of scholars who have mastered all the significant facts in their fields, often coupled with excellent ability to present such materials and illustrate them in the form of publications or lectures. All scientists collect and collate facts and make deliberate efforts to unify and interpret the various phenomena; many write textbooks. Yet there are only a few among them who also possess the ability of vision or insight needed to actually synthesize established knowledge into new concepts. Even those who possess such powers may have to wait long periods for flashes of intuition, which suddenly may illuminate puzzles that for some time had occupied their minds, seeming insoluble. This kind of productivity cannot be taught, and it cannot even be regulated by the individual himself. In relation to creativity, the man of science must be an artist as well as a scholar, and only after the gleam of insight arrives is he able to start to work on the details to formulate in acceptable terminology what originally appeared in his mind as a nebulous conceptual image, probably devoid of verbal content.

Characteristics of effective individuals

Several systematic studies of highly creative individuals have been reported in the literature. Lange-Eichbaum assembled biographic sketches on a great number of famous people from past times, attempting to describe some of their characteristics from recorded data. His books supply a large body of information as well as some analysis. Ghiselin collected additional biographic materials as well as descriptions by living subjects of their methods of work. Other authors, including Koestler, have concluded that the characteristics of highly creative persons in all fields are basically similar, that the creative process has definable roots, which involve unconscious as well as conscious mechanisms, the fundamental property being an exceptional ability to see relationships which are not apparent to the average person. The basic traits of the creative individual are thus the same irrespective of his field of endeavor, whether it be art, literature, music, philosophy, or science.

In a smaller but more detailed study of the actual life experiences of famous persons, Goertzel and Goertzel found that as a group they were seen as "different" by their contemporaries, that they tended to exhibit inferior performance in school, although there were exceptions, and that they were often characterized by poor social adjustment.

Still other studies have involved living individuals who were chosen by their peers as the most productive persons in their fields. With a battery of psychologic tests the personality makeup of such individuals could then be evaluated in considerable detail. Unconventional modes of thinking, including "overinclusion" as well as introverted tendencies, were among the characteristics of these groups, described in further detail by MacKinnon and by Barron. This approach has the advantage of dealing with living individuals, who can be subjected to specific studies of their personality makeup, but it obviously is limited to persons not necessarily of the very highest degree of creativity, since it is dependent on the judgment of contemporaries.

Individuals possessed of a creative impulse often seem compelled to limit their energies entirely to the problem at hand. With great persistence the chosen activity is pursued toward its goal, the fury of the creative drive and its constancy of purpose being all-consuming. This tenacity and single-mindedness, which may lead to an exclusion of family and social functions, has even been compared to the monomania of the lunatic.

In addition to the common characteristics which obviously have a bearing on their creative endeavors, men of genius share other traits, which do not appear to be directly allied to the productive effort. Celibacy or reproductive failure are very common. Stuttering or stammering, as well as bashfulness, have been established as frequent phenomena. Pallor, emaciation, and premature graying are mentioned by Lombroso. A tendency to social isolation or even alienation from their contemporaries has often been recorded. Great sincerity and a serious demeanor are said to be characteristic. Poor health is frequently encountered, but apparently this does not necessarily interfere with longevity. Some investigators have described a high energy level and vitality as typical, but certainly there have been many productive individuals who were not thus characterized. Some creative persons have considered themselves to by physical invalids, spending much of their lives confined to bed, yet able to remain productive.

Creative persons as a group thus appear to possess many characteristics which seem abnormal to the average person.

Psychologic tests of creativity

A different approach to the question of identifying potentially productive individuals has been based on a selection of accessible groups, often school children, on the basis of performance felt to reveal "creative" tendencies. Tests have been devised which detect unconventional thought patterns, abilities to come up with novel associations, or unusual quantities of responses. Some of these tests have been formulated on the basis of observed traits seen in recognized creative persons.

The fundamental problem in this method is that no one really knows whether such "creativity" tests actually measure the same qualities as those found in unquestionably productive persons from the past. Until some verification is obtained that the measured qualities indeed constitute true creativity, one must reserve judgment about the validity of this entire approach.

Some tests of creativity present the subject with a short story, asking him to record all thoughts that this stimulus evokes during a brief period. An alternate design is to employ a stimulus word, then tabulate all associations that this brings on. Still another approach is to present certain puzzles and note solutions that come to mind. Ratings are done in terms of fluency, novelty, or intuitiveness.

In reality the more conventional projective psychologic procedures, such as the Rorschach inkblot test, furnish a comparable standardized stimulus, to which the subject is asked to react, leading to scores which can be evaluated by comparison with known patterns of responses. Object-sorting tests have also been useful, again concentrating on the novelty of the productions. In either case, brain activity is turned on by a standardized stimulus, and the pattern of responses is evaluated against established criteria, which are based on previous studies of selected groups.

Obviously it is important to base the scoring of creativity tests on the responses known to be characteristic of recognized creative persons, rather than on what the experimenter himself prefers to see as indicative of creativity. It has, however, been the experience that individuals selected as creative on the basis of test performance do not exhibit quite the same traits as persons of proven creative productivity. For example, on the basis of test findings it has been claimed that "creative" individuals show good social adjustment, while studies of men of genius have generally led to the opposite conclusion.

If eventually tests should emerge which can be correlated with true creative behavior, a more clearcut definition will become possible, but until such time it seems safer to employ actual descriptions of recognized creative individuals as a valid guide. The basic qualities leading to creative production in any field, involving intuition, conceptual leaps, or novel associations, are thus what one has to think of as true creativity.

15
Evidence for Inheritance of Creativity

In view of the problem with a definition of the word "genius," it may seem strange that there is an impression that, in genetic terms, creativity may be a simpler concept to deal with than intelligence. However, since it appears that creative persons in all fields have in common a constellation of relatively simple characteristics, while the word intelligence carries with it no such unity, it can make sense that of the two, creativity perhaps has the simpler biologic basis.

To exhibit true creative intelligence an individual must possess superior learning ability as well as be endowed with the characteristics shared by all highly creative persons. There appear to exist men of creative capacity who lack an interest or ability in academic matters. Many scholars also seem to be missing the attributes which are needed for creative contributions in their fields. Creative scholars need not have distinguished themselves in terms of having been the top students in their classes, but obviously they must possess the qualities necessary for absorbing the information required to formulate important problems in a scholastic field.

Distribution of creativity

Since it is difficult to identify a large number of definitely creative individuals, there is no way to demonstrate with any degree of certainty whether creativity, defined according to the studies described in the previous chapter, generally runs in families. Some kinships have a cluster of productive persons, usually all working in the same field, such as the Bachs in music, the Bernoullis in mathematics, the Dumas in literature, and more recently the Bohrs in nuclear physics. The opinion is also often expressed that genius is born, not made. In most respects, however, the methods for genetic studies, discussed in Section I, cannot be applied directly to creativity as a concept, because of the limited family data.

Interestingly, there have been important regional differences in rates of creativity, the highest number of recognized productive persons having originated among Caucasian and Oriental populations. Galton described the abundance of productive individuals in ancient Greece. The reasons behind such clustering are, however, debated. The recognized chief contributors to any field of endeavor have for some reason always been males, leading to the postulation that masculine types of aggressiveness may be among the contributing factors.

No world-famous creative persons have been identical twins, perhaps in part because twins tend to suffer trauma in the intrauterine or early postnatal period, ending up with a somewhat lower IQ than the overall population. Twin studies of creativity are thus useless.

Effect of order of birth

One area of some interest is the observation that creative individuals are more likely to be firstborn or only children. Perhaps this ties in with recent findings, which have established that there is a small, but measurable, decline in intelligence with each successive pregnancy in a family. The reason for this latter phenomenon is at present unclear, but in view of the great regularity of the process, it seems likely that biologic factors play a role, possibly some kind of injurious process affecting later born children. Slight brain damage may blunt any creative propensities, thus explaining why the first child has an advantage and why twins appear to be at a disadvantage.

Although the increase in firstborn among creative persons has been used as an argument against heredity, this contention does not appear to have much validity. It may simply mean that a creative potential can be destroyed by adverse circumstances. However, the observed slant toward firstborn in the creative ranks certainly gives no support to hereditary influences.

Relation of creativity to mental illness

The above considerations still leave unsettled the question of whether creativity has any specific genetic basis, although there seems to be a consensus that productive tendencies must in some manner be acquired through heredity. Pedigree analysis with creativity as the criterion is of little use, and no quantitative family studies are possible. Data on twins or on foster-reared children appear worthless. In the midst of the dearth of evidence, however, one significant fact does emerge. It has throughout human history been observed that mental illness is common in creative persons, as in the adage of genius and madness being closely allied. Many scholars have resisted this concept and have tried to disprove such a relationship. However, the idea constantly resurfaces, taking on new dimensions.

One of the few systematic studies dealing with inheritance of creativity, *Hereditary Genius,* was published a century ago by the great scholar Francis Galton. He amassed a large body of data in support of his thesis, but present-day investigators feel that his material was slanted more toward giftedness than toward true creative capacity. However, his data do support the basic concept that genius is born, not made. Galton did not subscribe to the notion that genius may be related to insanity, although he eventually did observe that mental illness seemed unusually common in the creative ranks. Interestingly, he himself developed some kind of mental confusion in his later years, although no specific diagnosis was established. He is also said to have suffered a nervous breakdown, characterized as a depression, while he was a student at Cambridge University.

The studies reported by MacKinnon and his associates, which were mentioned earlier, dealt with creative architects and writers. The groups were selected on the basis of questionnaires distributed among scholars, asking them to nominate living individuals whom they considered the most creative contributors to these fields. Individuals who were most frequently nominated were invited to spend several days undergoing extensive tests, and thus a great deal of information was assembled. A finding that came as a surprise to the investigators was the similarity of the psychologic responses of these individuals to those familiar to all clinical psychologists as characteristic of psychotic patients, for example responses on the Minnesota Multiphasic Personality Inventory (MMPI) test. If the protocols had been handed to them as test results coming from hospitals, the high scores on psychopathology, including depression and introversion, would most likely have led to a diagnosis of schizophrenia. However, since they were known to be tests on creative persons, a more positive interpretation was seen as appropriate. Unconventionality, novelty, or ability to tolerate ambiguities were seen as desirable qualities in this group.

Another study, reported by Dudek, deals with Rorschach tests administered to eminent writers. The same theme recurs; responses ordinarily considered indicative of psychopathology are quite prominent.

In a third study, conducted by Andreasen, greater controls were added, and relatives of creative persons were also included. The index cases were well-known writers in some manner affiliated with the University of Iowa. This last investigation actually deals with the rates of diagnosed psychosis as well as with the psychologic characteristics of the individuals under study. Once more the findings indicate increased pathology in the ranks of famous persons. A significant increase in mental illness was also demonstrated in their relatives.

These findings open up a new field for exploration. The interest thus centers around the concept of a schizophrenia gene, whose primary function may in reality be in the creativity area, mental illness then being viewed as an unfortunate side effect. Besides helping to explain many findings from studies of creative persons, this concept can account for the high frequency of mental disorders in all human populations.

It thus seems that a promising area for investigation of the genetic basis of creativity impinges on the hereditary aspects of mental disorders.

16
Existence of a Schizophrenia Gene

Mental disease has always been known to have a familial tendency, although interpretations of this observation have been highly controversial. Because of the heated debates, proponents of the genetic view have eventually made thorough explorations along all possible avenues, and by now voluminous data have been gathered, using the various methods outlined earlier.

Before reviewing the extensive genetic data, it is necessary to describe the symptoms of psychosis and to touch briefly on the relationship between different forms of mental illness. Some investigators actually prefer to view all psychiatric problems as a continuum, with normalcy on one end of the spectrum and severe schizophrenia or dementia praecox on the opposite pole. In between the extremes these authors place personality disorders, neuroses, reactive psychoses, manic-depressive illness, and milder forms of schizophrenia. While the progression of severity of the disturbance can obviously be arranged in this fashion, there is no true indication that all mental disorders are biologically related or that they otherwise share the same etiologic basis. Scientific evidence for such an interpretation is nonexistent.

Signs and symptoms of mental disease

Full-blown mental illness usually does not make its appearance before adolescence, but the age of onset can span the entire period of adulthood. Psychiatrists generally agree that the disorders on the more severe end of the illness spectrum have common properties, and these are referred to collectively as the psychoses, subdivided principally into manic-depressive illness and schizophrenia. These disorders are in a different class than the generally milder neuroses. All psychoses lead to disturbances of sufficient severity to cause discomfort to the patient and interfere with normal social function, the degree of disability varying from somewhat aberrant behavior to total incapacity. Some of the common manifestations, such as hallucinations, are in reality neurologic or perceptual dysfunctions that would not be predicted as necessary concomitants of severe maladjustment.

The symptoms of psychosis have sometimes been divided into primary and secondary categories, the first including the general disorders of thought and affect and the second the various specific signs, such as paranoid delusions or hallucinations. The utility of this dichotomy is now questioned by many authorities, and its scientific validity is not established. In the majority of patients psychosis follows an episodic course, periods of illness being interspersed with intervals of normal health. Some individuals, however, develop mental illness in early adult life and remain permanently incapacitated.

Lack of enjoyment of life, sometimes referred to as anhedonia, is often a complaint among mental patients. Sleep disturbances are frequent, and disordered affect, such as a mood of euphoria or depression, is a very common occurrence. Other symptoms include manic behavior, flight of ideas, distortion of reality, or paranoid trends. Social withdrawal or inactivity are common. Grossly delusional thinking occurs in severe psychosis, and occasional patients are physically assaultive. The majority of psychotic patients appears to exhibit tendencies to social isolation, being quiet, unobtrusive, and nonparticipating. Suicide is frequent in their ranks, in both manic-depressive and schizophrenic patients. Failure to maintain acceptable standards of personal appearance is mostly seen in chronically sick individuals. Although males are more likely than females to end up as residents of institutions, systematic studies usually show 50 percent more total psychosis in females than in males, the excess being accounted for largely by depressive disorders.

Classification of the psychoses

Manic-depressive illness can be divided into the elated and depressed forms, with many patients following a circular pattern, showing alternating periods of euphoria and depression. Many such patients seem quite normal between the psychotic episodes, and typically there is no overt thought disorder.

Schizophrenia has also been classified into different subtypes, the four principal categories being paranoid, catatonic, hebephrenic, and simple schizophrenia. There is now a lessened interest in these subdivisions, as they are not always constant, and they contribute little by way of defining the approach to the illness. At present the custom in the United States is to diagnose most patients as either paranoid or chronic undifferentiated. A new category of schizo-affective schizophrenia also is often used, overlapping with manic-depressive psychosis.

The German physician Kraepelin was largely responsible for initiating the customary division of psychotic disorders into the two principal forms. He published his works almost a century ago, and while additional factors were involved, the separation was mainly based on the expected outcome or prognosis. Hence persons with a remitting form of psychosis were likely to be classified manic-depressive, while those manifesting a more severe illness, considered prone to be chronic, were diagnosed schizophrenic. The hallmark of distinction was delusional thinking, as milder manic-depressives may not exhibit overt delusions, while severe schizophrenics generally manifest quite disorganized thought patterns. However, it has been common practice to change the diagnostic label from manic-depressive to schizophrenic if the illness turned out worse than initially was anticipated, the true criterion thus in reality being the eventual outcome.

It is for reasons of this type that it has always been difficult to draw a definite line between the two main forms of psychosis, and diagnostic practices have varied greatly in different regions of the world. At present the majority of psychotic patients in the United States are classified schizophrenic, but in northern Europe a manic-depressive diagnosis remains more common.

There is much scientific evidence that the alternate forms of psychosis may in reality be different manifestations of the same biologic entity, thus justifying the American position of grouping them together under one diagnostic label. Overinclusive thinking, loose associations, or even delusional patterns are seen on psychologic tests in both manic-depressive and schizophrenic patients, and hallucinations as well as sleep disturbances occur in both. Elated or depressed states are not unusual in patients whom all authorities would consider schizophrenic. Both disorders also respond to the same forms of treatment.

Shields has shown that irrespective of the official diagnosis, an increased risk of psychosis is always seen in relatives of mental patients, including in each instance an increase in schizophrenia. The logic of considering the two principal entities separate disorders, rather than different forms of the same disease, is therefore not at all convincing. This analysis is not meant to deny that differences are truly demonstrable between classic forms of manic-depressive illness and schizophrenia. Instead it is intended to point out that they are overlapping disorders with many features in common, often showing progression of the disease from one toward the other.

Because of the impracticability of separating the affective versus schizophrenic psychoses for the purposes of a genetic study, the data presented below sometimes deal with all psychoses as a unit. Most of the data are, however, based on patients that generally would be considered schizophrenic, at least by American psychiatrists.

Family pedigrees of schizophrenia

Studies in several countries, including the United States, have established that psychosis afflicts more than five percent of the people at some time during their lives. No family is therefore free of psychosis, although some prefer to pretend to be spared. Large pedigrees have been published, usually starting with psychotic index cases, some spanning many generations. At times variable concentrations of cases are observed in different family segments, with both schizophrenia and manic-depressive illness being increased or decreased in the same branches.

In pedigree studies it is usually possible to document a certain degree of continuity of psychosis, but the pattern follows no simple Mendelian type of distribution. While clustering of affected members may be seen in certain areas of a pedigree, most cases of actually diagnosed psychosis are seen as isolated instances, mostly just one affected member in a family unit.

Although it is not possible to even suggest any particular genetic mechanism from a cursory examination of pedigree data, the overall impression is that the distribution is indeed consistent with a genetic type of transmission. Once psychosis appears in a family, it can be shown to persist, segregating to some extent into high and low rate branches. A typical pedigree, identifying documented patients, is shown in Figure 16-1 as an illustration of the usual pattern. If a comparable pedigree is assembled, showing definitely diagnosed psychotic cases in families not known to have psychosis in the initial generation, one commonly would see no fully documented case, or just one rather than several as shown here.

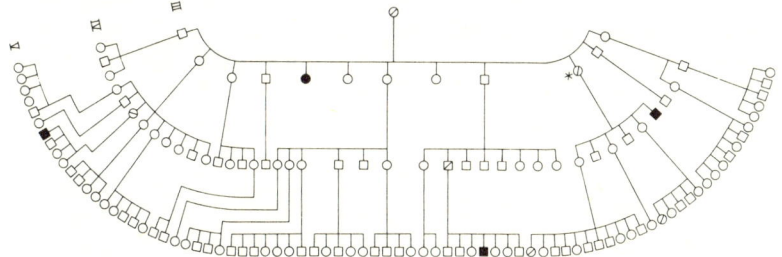

Figure 16-1. Pedigree of a family arising from the sister of a case with documented psychosis. Diagnosed mental patients are shown in black, parents of a psychotic person indicated by a diagonal line.

Quantitative family studies of schizophrenia

Many quantitative studies have been made of the risk of psychosis in families with known affected members. In general there is agreement that a definite increase exists, when a psychotic index case has been identified, but the studies have been performed in different ways, making a precise overall analysis of the total findings almost hopeless.

Among the more systematic family studies, most have followed the same general design. Usually this has consisted of identifying index cases associated with a particular mental hospital, selecting these according to preset criteria, and then attempting to identify and examine the relatives. All these studies suffer from the same fundamental inadequacies. For one thing, the relatives fall into different age groups. While methods of age correction have been devised, it may be difficult to make any meaningful comparison of mental illness, for example, in aged or dead grandparents versus young grandchildren. Obviously there are also immense problems of logistics when one attempts to diagnose via interview or history the problems of a large scattered population. Some of the studies have utilized a work force, which visited the relatives to gather diagnostic impressions.

Despite the serious problems, Zerbin-Rüdin has attempted to make an overall summary of all data of this kind. While one must view the exact figures with appropriate skepticism, the data in Table 16-1 are indeed quite impressive. There is obviously an increased risk in the close relatives, tapering off as one moves toward the more distant relations. The data of Kallmann, the largest single investigation, are also shown, and in reality the findings are essentially identical, except for Zerbin-Rüdin's figures being approximately one-third lower.

Table 16-1

Percentage rates of schizophrenia in relatives of schizophrenic patients, reported in previous family investigations

Relationship	Kallmann's data	Zerbin-Rüdin's summary
Child of two schizophrenics	68.1	46.3
Parent	9.2	6.3
Child	16.4	13.7
Sib	14.3	10.4
Half-sib	7.0	3.5
Grandparent	—	1.6
Grandchild	4.3	3.5
Uncle-aunt	—	3.6
Nephew-Niece	3.9	2.6
First cousin	—	3.5

The investigators who carried out the family studies usually did not perform population surveys of mental illness, these being furnished by other workers. Unfortunately, the two groups utilized different criteria, those doing population surveys diagnosing as schizophrenia only the most severely psychotic patients, preferring labels such as manic-depressive illness, psychotic-depressive reaction, or involutional melancholia for the less severely diseased. In the family studies, on the other hand, there was a tendency to assume that all psychotic relatives suffered from the same disease as the index case. For this reason one often reads in the literature that the risk of schizophrenia is increased fifteenfold in first-degree relatives, but this figure is erroneous. In studies where the same investigator has established the risk of psychosis in the general population and in close relatives, it consistently turns out that the increase in first-degree relatives is approximately fourfold. This holds true for schizophrenia, for manic-depressive illness, or for all functional psychoses combined, provided the criteria are the same for the general population and for relatives of index cases.

It will be noted that the established age-corrected risk in first-degree relatives tends to be in the order of 15 percent, with Zerbin-Rüdin's figures being somewhat lower than Kallmann's. The values for more distant relatives are considerably lower. The highest risk in these groups is seen in children of two psychotic parents, the best estimate being actually in the order of 35 percent.

Twin investigations

For the last half-century a controversy has raged over concordance rates for psychosis in twins. All investigators have arrived at the conclusion that the risk in monozygotic cotwins is approximately four times higher than that of dizygotic cotwins. There is also agreement that the psychosis is generally seen to be of the same type in both members of a monozygotic set, either manic-depressive or schizophrenic in both, and even the subtype of psychosis tends to be the same. However, when it comes to the exact magnitude of the concordance figures, the agreement comes to an end.

One problem in gathering twin data is that one ends up with pairs of variable ages. Obviously a discordant pair aged twenty years cannot be counted the same as a pair at age sixty years, as the former still has plenty of time to become concordant. Some authors have applied age correction procedures, but the heat of the arguments has been such that lately it has been considered unwise to utilize any corrections, although everyone agrees that some age adjustments are indeed in order.

Another problem has been the choice of sampling methods. The older studies were generally hospital-connected, leading to concordant twin sets being more likely to be encountered. Accordingly, the pair method of computation of concordance rates was used, as is appropriate for this type of material. In more recent studies attempts were made to locate and study all existing twins in selected populations, leading to a form of total ascertainment. Unfortunately, the authors of these latter studies, not being geneticists, were unfamiliar with the proper procedures and thus applied the wrong calculation methods to their data. This resulted in concordance rates only one-half as great as the previous figures, causing a flurry of excitement and claims that the old figures had been over-inflated.

Concordance rates for schizophrenic illness in monozygotic twins, reported by various investigators, are summarized in Table 16-2. These figures are not corrected for age, so that the true rates are likely to be higher. The first group shows the hospital-based studies, and the concordance is computed with the pair method, which is appropriate for this type of sample. The rates vary from 54 to 71 percent when milder cases are included in the data.

Table 16-2

Concordance rates for schizophrenic illness in monozygotic twins, without age correction, lower value showing rate for strict schizophrenia, upper value inclusive of borderline cases

Investigator	Number of pairs	Pair method concordance %	Proband method concordance %
Hospital-based studies			
Luxenburger	14	50-71	
Rosanoff et al.	41	44-61	
Essen-Möller	7	14-71	
Kallmann	174	59-69	
Slater	37	49-65	
Inouye	55	36-60	
Gottesman and Shields	24	42-54	
Population-survey studies			
Tienari	14	(14-36)	25-53
Kringlen	55	(25-38)	41-55
Fischer et al.	21	(24-48)	36-56
Pollin et al.	95	(27-)	43-

The second group of studies covers the population surveys, in which all twin pairs with psychosis in a given area or a selected subgroup are included. The pair concordance for these samples is shown within brackets, since this method is not applicable to such samples. When the proband method is used, the rates in these latter studies become comparable to the older data.

Twin studies of manic-depressive psychosis have also been reported, and the outcome is very similar to the results with schizophrenic pairs, two-thirds being concordant for manic-depressive illness.

At this point there is still no overall agreement in this field, but most investigators are willing to concede that if one monozygotic twin suffers from a psychotic illness, the risk is at least 50 percent that the other twin will also succumb. Kallmann, who did the largest studies, actually claimed 85 percent concordance for schizophrenia, but many authors now feel that this figure is too high. As was previously stated, the concordance rate in dizygotic twins is generally only one-fourth as high, being the same as the risk observed in regular sibs.

Since dizygotic twins share environmental conditions much more closely than regular sibs, while the genetic relatedness is the same, the fact that the risk is no higher in dizygotic cotwins than in regular sibs seems to contradict the hypothesis that the family environment is the major factor leading to psychosis. Another interesting observation, reported from Denmark, is that the risk of psychosis in the children of twins with psychosis is just as high whether one starts out with schizophrenic twins or with their discordant monozygotic cotwins. In this approach one is in reality studying the genes of the twins rather than the symptoms shown by the individual.

Additional studies have described monozygotic triplet brothers, all schizophrenic, and a set of monozygotic quadruplet sisters, all suffering from a catatonic form of schizophrenia.

Foster-reared relatives and schizophrenia

The controversy over the family and twin investigations has recently led to a renewed interest in individuals reared away from their biologic families. Such data are not easy to obtain, since it is necessary to deal with adult subjects, information about their childhood experiences being difficult to document.

As an illustration, it seems useful to give an account of the experience with such a study done in connection with the present research. Having just returned in 1964 from Iceland, after spending a year gathering data on psychotic families, as well as limited information on foster-reared individuals, the writer surveyed the records at the Napa State Hospital in California to locate any individual born to a psychotic mother subsequent to her admission with a schizophrenic diagnosis. The material was divided into three fifteen-year periods, identifying a total of thirty-four potential cases. Only eleven were found to have been placed in foster homes. Since it was considered essential to ensure early separation from all biologic relatives of the mother, the records were carefully screened and all cases eliminated where a significant contact was documented between the child and the mother's family, including her parents, sibs, and previously-born children. It turns out that social workers make great efforts to maintain family contacts, so that only five of these children had experienced total separation from their relatives. Furthermore, of the remaining group, two had been adopted, and in California such legal barriers have been erected against the identification of adoptees that it is hardly worth trying. However, the adoptive parents were identified in one instance. In the end just four satisfactory cases thus resulted from this time-consuming survey.

With the benefit of this experience a new design was developed, this time identifying all women admitted during a selected period who had given birth to their only child shortly before admission to the hospital. These children were unlikely to have been contaminated by contact with the mother's relatives; in fact, it was permissible for them to be reared by the father. Limiting the study to mothers who either died or remained continuously hospitalized at least five years, eliminating also a few children who were reared by maternal relatives or placed in adoptive homes that could not be identified, a total of eleven cases were located. To this were added three firstborn children from the previous survey who happened to fit into the appropriate time interval. Unfortunately, funds could not be obtained for a complete follow-up of these 14 individuals, since opponents of the genetic view generally hold the pursestrings and are reluctant to support studies which threaten their position. However, a survey of California mental hospitals showed that three had undergone treatment, all with schizophrenic diagnoses. One additional separated case was born to a psychotic father, and that daughter was herself healthy, although she in turn has had a psychotic child.

Considering the difficulties encountered in this type of investigation, it is not surprising that the total information of this sort is still rather limited. However, when one adds to the above data (15 cases, 3 schizophrenic) the findings of Heston, published in 1966, (47 cases, 5 schizophrenic), the Danish study published by Rosenthal in 1972 (44 cases, 3 schizophrenic), and a recently reported series by Cunningham and his associates (8 cases, 1 schizophrenic) the total material comes to 114 with 12 diagnosed schizophrenic. Two of these studies, which were carefully controlled, provided data on 117 children born to non-schizophrenic mothers, also reared in foster homes from early infancy, and none of them has been actually diagnosed schizophenic, although a few cases are reported to show borderline features.

Environmentalists, while willing to concede that these data seem impressive, have brought up the claim that this type of study does not rule out intrauterine factors. This has been countered with the argument that the psychosis risk in children of schizophrenics was established a long time ago by Kallmann to be approximately 10 percent uncorrected or 15 percent corrected, whether the affected parent was the father or the mother. This makes it difficult to maintain that intrauterine factors are paramount. Besides, there are reported twelve sets of monozygotic twins with schizophrenia who were separated early in life, nine of these being concordant. Here there are no significant intrauterine emotional differences, and it is not even possible to maintain that the second twin may have been influenced early in life by being told about psychotic relatives, since the index twin was still healthy during that period.

A few studies have involved still other designs. For example, data were gathered in Iceland on foster-reared individuals who became psychotic in adult life, comparing the rate of psychosis in their biologic and foster relatives. The eight index cases had twenty-eight foster sibs, none of them psychotic, while the twenty-nine biologic sibs, reared elsewhere, showed six instances of schizophrenia. A study was also made of sibs of schizophrenic index cases who had a psychotic parent, indicating no reduction in the risk of psychosis in a subgroup that was reared in foster homes from early infancy. The various studies of separated relatives are summarized in Table 16-3, comparing the observed data with predictions based on environmental versus genetic hypotheses. The risks in various relatives obviously are in good agreement with the genetic model.

Table 16-3

Summary of data on schizophrenia in foster-reared relatives of chronic mental patients, showing number of schizophrenic cases among individuals separated from their families before one year of age

Type of relationship	Number studied	Expectation of schizophrenia		
		Environmental theory[1]	Genetic theory[2]	Cumulative data
Monozygotic cotwins	12	1	8	9
Children of two schizophrenics	2	0	1	1
Sibs, one parent affected	9	0	2	3
Children of a schizophrenic	114	6	13	12

[1]Number of cases, assuming 5 percent rate of psychosis for population of average risk
[2]Based on Kallmann's uncorrected findings for various relatives

It has been considered virtuous to accept the argument that schizophrenia most likely results from an interaction of both genetic and environmental factors, and one often hears statements to this effect from individuals trying to follow a conciliatory course. However, since the risk appears to be equally high with or without separation in children born to psychotic parents, it is hard to see that the actual data give much support to environmental etiology. Everyone recognizes that genes do not operate in a vacuum, but one is left with the conclusion that the risk of schizophrenic illness appears to be already largely fixed at the time of conception.

The skeptics in this field have been quite willing to use scientific arguments when it served their purpose, but now that the appropriate data have been supplied they appear to be disinterested and still unconvinced. However, it does not seem excessive to claim that the operation of genetic factors in schizophrenia has been established. The data simply speak for themselves.

17
Genetic Transmission of Psychotic Tendency

Accepting that the evidence from studies of twins and foster-reared individuals, coupled with what already was known about increased family risks, indeed has established that genetic factors are operative in schizophrenia, one can proceed to the next step of trying to decide what type of genetic mechanism is involved. It is recognized that none of the ordinary simple genetic systems is capable of accounting for the family data.

Many different modes of inheritance have been proposed, including on the one hand modified versions of all possible one- or two-gene combinations and on the other hand polygenic mechanisms of various sorts. Data have now become available which permit a comparative assessment of these various models.

Evidence against polygenic inheritance

Although it has become fashionable in recent years to invoke polygenic explanations for any genetic disorder which a priori seems to fail to fit one of the standard simple types of inheritance, the fact remains that this kind of transmission has never been established to be responsible for any human disease.

One effective approach to an evaluation of a polygenic hypothesis can be based on a consideration of the necessary gene frequencies. Since a schizophrenic type of illness occurs in over 1 percent of the population, perhaps even in over 5 percent, the individual members of a polygenic system obviously have to exist at very high frequencies. Several infrequent genes cannot occur together in a significant proportion of the population. Even just two genes, each carried by as high a proportion as one-tenth of all persons, exist together in only 1 percent of the people. Furthermore, a polygenic system proposed in this type of setting deals with nondetrimental genes, the unfavorable combination being the source of the disorder, rather than abnormal gene action. It thus becomes merely a matter of chance when such a combination occurs, all kinships being equally vulnerable, since frequent genes exist everywhere. From this it is evident that a polygenic mechanism should result in roughly equal frequencies of the disorder, not only in all kindreds, but in all branches of any large kindred.

Longitudinal data have been presented on a large kinship studied in Iceland, into which the first documented case of psychosis was born in the year 1735. A comparison of the occurrence of recorded psychosis in the six branches of descendants derived from the original ancestors is illustrated in Figure 17-1. It is apparent from inspection of the figure that psychosis is quite unequally distributed in the various segments. Branch A has twenty-two documented cases in generations I to VI, while just one case has appeared in the last generation in branch C. Furthermore, in generations VI and VII, which arose after medical diagnoses had become customary, branch A actually has ten documented cases of schizophrenia and eleven cases of affective illness, while in these same generations branch C has just two cases of schizophrenia and three of manic or psychotic depression. Besides the findings within this kindred, there is considerable variation in the rate of recorded psychosis between different kinships in Iceland.

These data appear to be inconsistent with polygenic inheritance and instead suggest a simpler mechanism, which can permit the observed type of segregation into high-rate and low-rate segments. Modified versions of one or two gene systems at this point seem more attractive.

Distribution of psychosis within families

As was stated above, inspection of family pedigrees eliminates

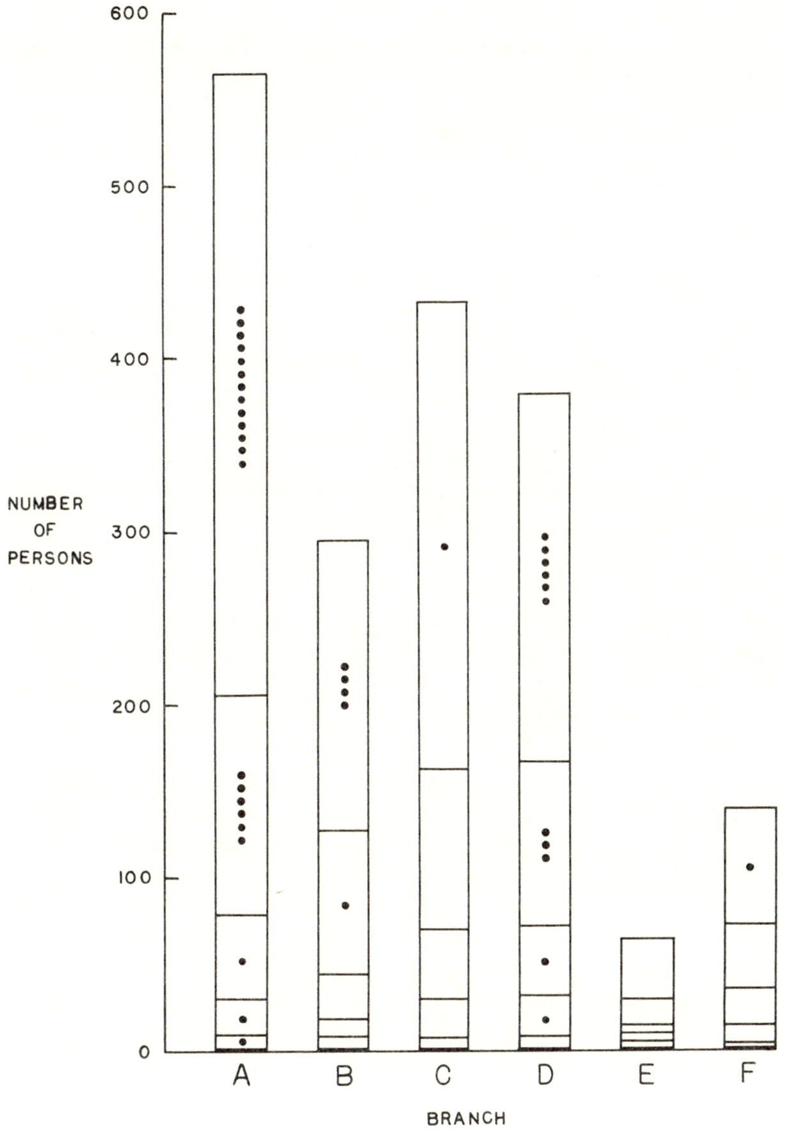

Figure 17-1. Distribution of psychosis in the six branches descended from a couple born in Iceland in 1682. Each column starts with one of the children of the original couple, with five additional generations placed atop each other. Black dots denote cases of documented psychosis.

any unmodified single-gene mechanisms. Schizophrenia does not generally pass from an affected parent to one-half of the offspring, nor does it appear in one-fourth of the sibs when both parents are unaffected. No form of sex-linked transmission needs to be considered, as the distribution follows no such pattern. Even a modified form of recessive inheritance would require astronomical gene frequencies to explain the data. Furthermore, since the risk is very high in the children of two psychotic parents, it appears unlikely that there exist several genetically distinct forms of schizophrenia. The concept of balanced polymorphism also favors a unitary mechanism. One is thus left with some form of modified dominant autosomal transmission, which conceivably can involve an interaction of more than one genetic locus, as the only plausible mechanism that needs to be evaluated in greater detail.

The older family data, summarized in Table 16-1, have been used in many attempts to analyze the genetic mechanism, but more recent information indicates that these figures may contain serious systematic errors. Newer data, gathered by this writer on the population of Iceland, are shown in Table 17-1. Because of great population stability and a long-term interest in genealogy, as well as a high level of literacy and excellent medical standards, this country of just over two hundred thousand inhabitants is ideally suited to genetic studies of frequent disorders. Even the fact that women retain their maiden names after marriage works to the advantage of investigators trying to trace long-term relationships. The Icelandic data indicate that the older figures showed too low rates in the more distant relatives, but otherwise the new findings are in harmony with reports by other investigators.

Since the details of the family data from Iceland differ from the older findings, it is important to compare the methods and evaluate the design in the two types of investigations. As was pointed out before, the older data were generally gathered by one overall design, although the diagnostic opinions naturally varied between different investigators, leading to a fluctuation in the reported rates. Most started out with a hospital population, attempting to identify and examine the relatives. All these studies thus ended up with relatives in variable age brackets, and corrections had to be applied to make the figures comparable. The data in Table 16-1 are thus all age-adjusted values.

Table 17-1

Rates of psychosis in relatives of psychotic index cases in Iceland

Relationship	Number studied	Comparative rate of psychosis	Total risk of psychosis
Parents	1056	3.8	15.2
Sibs	2315	3.8	15.2
Children	393	5.0	20.0
Uncles-aunts	1093	2.2	8.8
Nephews-nieces	1006	2.3	9.2
First cousins	2332	1.3	5.2
General population	24171	1.0	4.0

To bypass the need for such corrections, which always lead to disputed figures, the Icelandic study was so designed that the ages of the index cases were allowed to vary, maintaining the ages of the relatives constant. For this purpose the probands were divided into three generations, born in the intervals 1851-1880, 1881-1910, and 1911-1940. The diagnosis made by the medical staff of the Kleppur Mental Hospital in Reykjavik, which serves all of Iceland, was accepted, dividing psychotic patients into those considered schizophrenic, manic-depressive, or psychotic-depressive, the last category being a combination of several related syndromes. While some investigators prefer to make their own diagnoses, this is unnecessary in a comparison of large groups, as the errors are unlikely to be clustered.

In the main study the relatives were confined to those born in the interval 1881-1910. This included children and nephews-nieces of the oldest generation, sibs and cousins of the middle generation, and parents as well as uncles-aunts of the youngest generation of index cases. A variety of different types of relatives was thus available, and no age correction is needed. The different groups were then compared in their mental hospital admission rates for psychotic disorders, the figures being expressed in multiples of the already established rates for the general population. The comparative rates of psychosis thus derived, shown in the middle column of Table 17-1, could be converted into total psychosis rates by the use of data on psychosis in persons born in Iceland between 1895 and

1897, which Helgason had previously published, basing his figures on a total followup for a lifetime of a selected cohort of Icelanders. Table 17-1 shows only the findings on all psychosis combined in the relatives of all psychotic index cases, but the comparative rates are essentially the same if one limits the data to the rates of schizophrenia in relatives of schizophrenic probands.

Just as others have reported, the total psychosis in females was found to be higher than in males, and was largely accounted for by an increase in depressive disorders. Since males are more likely to create social problems or remain economically dependent on social agencies, they are, however, generally overrepresented as residents in institutional populations. The risks of psychosis in the relatives in the Icelandic study are essentially the same whether one starts with male or female groups of index cases.

Since the Icelandic study is based on large numbers of relatives, utilizing a sound methodology and a uniform system of diagnosis, the resultant figures can be seen as quite meaningful. Reliable estimates of family risks make it possible to evaluate further whether modified forms of dominant inheritance indeed can account for the distribution.

Modified dominant inheritance of schizophrenia

As was mentioned earlier, interpretations of the older family data, which are listed in Table 16-1, have led to various proposals for a genetic mechanism for schizophrenia. Kallmann was for many years the chief authority in this field, having contributed the largest quantity of data. Early in his career he opted for modified recessive inheritance as the most plausible mechanism, and he stuck to that interpretation till his death, even when others could see serious contradictions.

Böök gathered more limited information on an isolated Swedish population, and while his data in reality were insufficient to strongly support any specific hypothesis, he had the benefit of greater experience in the field of genetics. Intuitively he favored some sort of dominant transmission and thus proposed a system based on a single major gene, illness occurring in approximately one-fifth of all heterozygous carriers and in all homozygous individuals. In his own material Böök seems to have classified most psychoses as schizophrenia, thus arriving at a rather high rate of the disorder in the general population. He consequently found only a fourfold increase in the first degree relatives, employing for them the same standards as for the population at large. Böök's proposal was later applied to the world data by Slater, with the conclusion that it had considerable merit, and it has remained one of the viable hypotheses for the last quarter century.

In connection with the earlier phases of the Icelandic investigation, the present writer attempted to interpret limited new information as well as the total world data, coming up with a scheme requiring a dominant principal gene, acting together with a gene at a different locus following a recessive mode of transmission. This proposal appeared to fit nicely the data of Kallmann, as well as agreeing with Zerbin-Rüdin's overall summary. The fit to the older world data was considerably better with this mechanism than with Böök's scheme.

In reality the purpose of the newer extensive family investigation, conducted more recently in Iceland and summarized in Table 17-1, was to attempt to provide more convincing evidence for the proposed two-locus hypothesis. As it turned out, the new data fit better the old interpretation elaborated by Böök. The crucial finding is that the rate in first-degree relatives is increased only fourfold over that of the general population, rather than fifteenfold as previously had been maintained by most authorities. The risk in more distant relatives also turns out to be higher than was reported in the older investigations. The difference presumably arises from the fact that there was a tendency in the past to classify all psychosis in close relatives as schizophrenia, while in other relatives and in the general population psychotic illnesses were partitioned into the different categories, only some being considered schizophrenic.

From Böök's scheme it is possible to derive a rather simple formula to express the expected risk in various relatives of index cases. In the case of the Icelandic population, the data suggest that the schizophrenia gene occurs at close to 10 percent frequency, one-sixth (17 percent) of the heterozygous carriers developing psychosis at some time, and all the homozygotes becoming mentally ill. One thus ends up with the expression $1.0 + 6.0 \, (1/2)^n$ as the predicted comparative risk of psychosis in different relatives in terms of the population rate, or close to $4 + 24 \, (1/2)^n$ as the total risk in various groups, n being the number of genetic steps removed from a psychotic index case (respectively, 1, 2, or 3 for first-, second-, or third-degree relatives). These simple formulations enable one to predict the findings from the Icelandic study, which probably represent more precise risk figures than the older data. The expected risk in various groups is then in the order of 16 percent for first-degree, 10 percent for second-degree, and 7 percent for third-degree relatives, as compared to a 4 percent lifetime risk of psychosis in the general population. Obviously these figures are in satisfactory agreement with the data in Table 17-1.

According to the best-fitting hypothesis, there thus exists a mutant gene, designated s, which occurs in the general population at a frequency of 0.1. Utilizing the Hardy-Weinberg law, 81 percent of all persons are noncarriers, SS, approximately 18 percent carry one mutant gene, designated Ss, and 1 percent of all persons are homozygous, ss. The noncarriers are free of psychosis, but one-sixth of the Ss carriers, accounting for 3 percent of all persons, develops a psychotic illness, mostly of the late-onset and remitting type. All the abnormal homozygotes become mentally ill, and presumably these illnesses include the severe type of psychosis that psychiatrists sometimes refer to as process schizophrenia. Thus a total of 4 percent of the population develops some form of functional psychosis.

This hypothesis fits the new family data, and it is an entirely plausible genetic model. It can be classified as a form of dominance with incomplete penetrance, encompassing the additional feature that the abnormal homozygotes must be included in the scheme, because of the high frequency of the mutant gene.

Opposition to the genetic evidence

Although the foregoing discussion has been presented in a manner which perhaps creates the impression that the data are now rather conclusive, it should be made clear that this is not at all the view in official circles. The scientific evidence has been available for some time, but the interpretations discussed here are not those favored by scholars in charge of psychiatric establishments. Young psychiatrists or psychologists in training are generally not introduced to these developments, which after all are in conflict with prevailing theories.

Notwithstanding this negative official attitude, people generally are under the impression that public funds allocated to pay for research and training are used wisely and made available to encourage promising explorations. Few would be so cynical as to suggest that such resources could be utilized to suppress new knowledge. Everyone anticipates jubilation in the scientific world when important problems are solved, expecting praise to be heaped on the investigators who contribute to a triumph over a human disease.

Perhaps the official response to the research findings in schizophrenia, including the rejection of Böök's brilliant deductions, is a measure of the fallacy of this public conception. There has been no rejoicing over the developments described here, just silence or overt hostility. The National Institute of Mental Health still has no Genetics Laboratory, although there is a Psychology Laboratory and a Sociology Laboratory. When a special Center for Studies of Schizophrenia was funded by Congress several years ago, an antigeneticist was appointed as its head. No funds could be obtained to carry out the research on the Icelandic population described in this book. Even the *Journal of the American Medical Association* refuses to print the new data.

Opponents of the genetic view have been supported liberally by public funds. The authors of newer twin studies of schizophrenia, who claimed to have disproven the importance of heredity, were flown all over the world to spread the gospel, even though a close look at their data made it apparent that the findings were arrived at by erroneous methods. The views expressed in this book, although based on far more information and well supported by various kinds of solid evidence, have been received differently. Such are the ways of administrators and even some professionals.

18
Relation of Creativity to the Schizophrenia Gene

If psychotic symptoms can be assumed to be caused by a chemical aberration, occurring in individuals born with a specific schizophrenic constitution, it becomes easy to visualize a relationship between psychosis and creativity. There is agreement now that overarousal appears to be a constant feature of schizophrenia, manifesting itself even in individuals who on the surface seem withdrawn or depressed. Evidence of this has been obtained on various physiologic measures, including the electroencephalogram. Conditions that increase arousal lead to rapid, low voltage electrical activity in the brain, while a reduction in arousal causes slower, high voltage activity, this being also seen during rest or sleep.

In a recent report, Mednick and his group, who are following the developmental progress of Danish children born to schizophrenic mothers, have demonstrated decreased latency in the evoked potentials measured in the brain responses of these individuals by refined electroencephalography. This is interpreted as further evidence of a state of overarousal, associated with the schizophrenic constitution. They have also utilized other physiologic techniques, such as measurements of the galvanic skin conductance, to demonstrate hyperreactivity.

Arousal is known to be intimately linked to the reticular activating system, located in the area of the brain stem and extending to the midbrain. This system initiates electric impulses which then travel to the higher centers, stimulating their activity. One can postulate that overarousal may lead to creativity, but in persons unable to tolerate the excessive stimulation, psychotic symptoms may appear.

Enhancement of brain function in schizophrenia carriers

Given a specific gene associated with schizophrenia, one can conclude from the calculations that this factor is carried in a single or double dose by almost 20 percent of the population. The majority of carriers never suffers from mental disease. Attempts obviously need to be made to assess what psysiologic effect the schizophrenic constitution may have on brain activity.

Because of the striking thought disruption in severe schizophrenic illness, an interference which at times prevents all logic or reasoning, investigators have in the past tended to concentrate on causative factors presumed to have their locus of action in the higher centers of the brain. Until recently, with environmental theories preeminent, the popular approaches have focused on early personality development, postulating aberrant input in such forms as improper mothering, early sexual conflicts, or deviant intrafamily communication. Of course psychiatrists have also been aware all along of the many other disturbances in schizophrenia, including malfunctions of the autonomic nervous system, but the disorder of thought has always been seen as paramount.

In view of these commonly endorsed interpretations, it comes as somewhat of a surprise that newer findings, instead of locating the basic disorder in the higher centers of the brain, actually seem to place the chemical aberration of schizophrenia in the most primitive portions of that organ, the part that man shares with fishes and other lower vertebrates. In the mammals, and in particular in man, the cerebral cortex—which contains the centers of association and reasoning, in addition to regulating the voluntary control over many functions—has so overgrown the lower areas that the latter occupy only a small region in the center of the human organ. However, that center region, operating in part through the so-called limbic system, still plays an important role in the adjustment of basic body functions, in the governing of emotional states, and in the regulation of levels of arousal. Malfunction in this region is capable of upsetting the entire machinery, and superior operation conceivably can improve on all brain functions, including intellectual performance. On the other hand, it is hard to imagine how inadequacies in the early environment, such as psychic trauma in infancy, can significantly affect the operation of this primitive and largely autonomous region of the brain.

It is not difficult to visualize a relation of the lower center arousal mechanisms to creative and learning processes. Psychologists have consistently found that states of arousal and motivation are vital components of the learning experience and that emotional content influences retention and memory. If one proposes that highly creative persons must carry the schizophrenia gene, this leads to the expectation that their responses on psychologic tests should be similar to those of psychotic patients. This formulation can also explain the high rate of psychosis observed in creative persons and their relatives.

While the most convincingly demonstrable positive effect of the schizophrenia gene thus appears to be its influence on creativity, there are some signs that this gene may contribute to scholastic performance as well. Up to this time there are no data indicating that as a group individuals who later develop psychosis perform better than others in school, prior to their illness. In fact, the evidence seems to point in the other direction, although it is somewhat difficult to interpret. With many such individuals experiencing in childhood adverse effects of the incipient illness, this may inhibit their attainment in school. One study, limited in scope, suggests that individuals later destined to become manic depressives may before that perform above average on IQ tests.

When one examines this matter further, limiting the study to top-level students, there are indeed indications that they and their relatives may have an increased risk of psychosis. This was the finding in a study of honor students in the Icelandic population, showing a higher psychosis risk in their ranks than in the population at large. The much larger Terman study of gifted students in California is not as clear on this point, as it lacks a control group, but a high rate of suicide was encountered.

If top-level school performance indeed is associated with the schizophrenia gene, it is possible to argue that this need not necessarily reflect improved learning, but rather is related to creativity. A very capable student, lacking in creativity, may be at a disadvantage when competing with a similarly bright colleague who benefits from creative thinking. For example, this may show up in abilities to solve difficult mathematical problems. Creativity may thus enter the picture when one attempts to compare the performance of top scholars, although only a fraction of all creative persons enter that arena.

It is of interest that the gene responsible for myopia, in the process of achieving a new balance because of altered conditions, shows great variation in frequency between different areas. In contrast, the schizophrenia gene appears to exhibit a more even distribution, being fairly frequent in all regions of the world and apparently in a state of equilibrium. No place is known to show very excessive rates of psychosis, and neither is any area free of such disorders. Perhaps a population must include some inventive members in order to emerge from primitive existence, and groups with a certain increase in schizophrenia genes have been the only ones to survive and multiply. Creativity may then be seen as a factor in group adaptation, only a few such persons being needed to insure the success of a population. Highly inventive persons, who alter the course of history, occur at a very low rate, but when their products become available, an entire population benefits. That is exactly what happened when a few individuals developed the use of electric current, and the experience was later repeated with radio transmission. Even military superiority in the present-day world is entirely dependent on a few highly creative minds.

Historical association of creativity with psychosis

In ancient times people often equated creativity with a supernatural or divine force, at times believing that a man of genius was possessed of a spirit. These theories were never actually disproven, but rather were bypassed by a shift of interests, few scholars any longer believing in the reality of such notions.

The early Greeks were struck by the frequent occurrence of insanity among highly creative men, and one reads comments like "men illustrious in poetry, politics, and arts have often been melancholic and mad," or "delirium can take two forms, either ordinary madness or heavenly inspiration or exaltation."

During the middle ages similar thoughts persisted, describing how great wit and madness were closely allied. Somber and melancholic temperament, often seen in creative persons, was ascribed to a periodic derangement of the organism.

The revival of interest during the last century in the frequent similarity of inspiration or ecstasy to madness has met with criticism, some authors feeling that the purpose of such analysis is belittlement of creative persons. During the second half of the nineteenth century several authors published accounts of the concurrence of genius and insanity, culminating in Lombroso's *The Man of Genius,* published in 1891 and widely circulated. For some time it has been customary to castigate such works and ascribe their production to a perennial distrust of anything novel or original or to hatred or assaultiveness toward those who seem to be above the masses.

The anecdotal approach to the relationship of creativity to psychosis has been replaced by more systematic studies during the present century. Kretschmer published a book describing many of the personality characteristics of recognized creative individuals, including some information about their mental health. This was followed by Lange-Eichbaum's very extensive survey of famous men, which still serves as a resource for those interested in the factual data. An analysis of his material led to the conclusion that psychosis indeed is frequent among men of creative genius, perhaps affecting a third of those of the highest creative caliber. Several other studies, such as those of Nisbet and Ellis, have further supported a relationship of genius to insanity. Representing the opposing viewpoint, Juda claimed to find only a slight increase in psychosis in a large survey of eminent Germans, but her study included a number of officials and scholars, whom few would rank among the highly creative. Interestingly, she reported that when they exhibit psychosis, scholars and scientists tend to fall into the affective categories, whereas poets and artists are more prone to develop delusional psychosis.

As part of the present research, a survey was made of famous individuals who were born in the United States west of the Mississippi River before 1890. Since the population of that area was then still relatively small, the total number of men of world fame was found to be rather limited. The ten individuals so identified were Lee De Forest (inventor), T. S. Eliot (poet), Robert Frost (poet), Edwin Hubble (astronomer), Sinclair Lewis (novelist), Jack London (novelist), Edgar Lee Masters (poet), Ezra Pound (poet), Josiah Royce (philosopher), and Mark Twain (humorist). In this group it is reported that T. S. Eliot suffered a nervous breakdown at age thirty three, Jack London exhibited manic and depressive episodes before committing suicide at age forty, and Ezra Pound was diagnosed as schizophrenic in his mid-fifties. In addition, it is recorded that Lee De Forest and Robert Frost, although apparently never considered psychotic, had episodes of moderately severe depression. Frost's sister and his children suffered from mental illness. Edgar Lee Masters had some kind of nervous exhaustion after writing his most famous poems. Mark Twain also became rather melancholic in his later years, and his children exhibited mental disorders.

A different survey of contributors to creative endeavors was conducted through identification of famous persons listed in various biographic compendia. Information on the mental health of the individuals chosen for inclusion in such books was obtained partly from other sources. This study is summarized in Table 18-1. There may well be debate whether all those considered psychotic are properly classified, and some who are not so listed are known to have been far from normal in behavior. The opinions accepted here are the best information that could be culled from the available literature. Psychosis obviously abounds in these groups, being just as frequent as Lange-Eichbaum reported in his earlier study.

These surveys appear to confirm once more previous impressions that highly creative men have an increased rate of mental illness. The various quantitative studies, which have limited the investigative material to individuals unquestionably of the highest creative caliber, show an impressive consistency in regard to this finding.

Creativity in the ranks of psychotic persons

For the general population, the lifetime risk of ever developing psychosis is estimated to be in the order of 5 percent, and certainly under 10 percent. It thus becomes possible to assess whether those individuals who become mentally ill are truly as a group a burden on society, a role they have often been assigned in the past. As was mentioned before, it is generally recognized that human progress has been dependent on a few great men, who have produced the inventions on which cultural advancement rests. Such individuals oc-

Relation of Creativity to the Schizophrenia Gene

cur at a rate of less than one in a million, and one crucial question is whether persons with a history of psychosis exhibit a still lower rate of creative productivity.

Table 18-1
Survey of famous persons listed in biographic sources on creativity, showing in italics individuals stated to have developed psychotic disorders

Source book	Persons listed	Rate of psychosis
Living Biographies of Famous Novelists		6/20
Boccaccio, Rabelais, Cervantes, Defoe, *Swift,* Sterne, *Scott,* Balzac, Dumas, *Hugo,* Flaubert, Hawthorne, Thackeray, Dickens, *Dostoyevsky, Tolstoy, Maupassant,* Zola, Twain, Hardy		
Living Biographies of Great Poets		6/20
Dante, Chaucer, Villon, Milton, *Pope, Burns,* Wordsworth, *Coleridge, Byron, Shelley,* Keats, Browning, Tennyson, Swinburne, Bryant, *Poe,* Longfellow, Whittier, Whitman, Kipling		
Living Biographies of Great Painters		7/20
Giotto, Michelangelo, *Raphael, Da Vinci,* Titian, Rubens, *Rembrandt, El Greco,* Velasquez, Hogarth, Reynolds, *Turner, Goya,* Corot, Millet, *Van Gogh,* Whistler, Renoir, Cezanne, Homer		
Living Biographies of Great Composers		7/20
Bach, *Handel,* Haydn, Mozart, Beethoven, *Schubert, Mendelssohn,* Chopin, *Schumann,* Liszt, Wagner, Verdi, Gounod, Brahms, *Tchaikovsky, Rimsky-Korsakov,* Debussy, Pucchini, Sibelius, Stravinsky		
Men of Mathematics		8/32
Descartes, Fermat, *Pascal, Newton,* Leibniz, J. Bernoulli, N. Bernoulli, Euler, *Legrange,* Laplace, Monge, Fourier, Poncelet, Gauss, Cauchy, Lobatchewsky, Abel, *Jacobi, Hamilton,* Galois, Sylvester, Cayley, *Weierstrass,* Kowalewski, Boole, Hermite, Kronecker, Riemann, Kummer, Dedekind, Poincaré, *Cantor*		
Living Biographies of Great Philosophers		8/20
Plato, Aristotle, Epicurus, Aurelius, Aquinas, Bacon, *Descartes,* Spinosa, Locke, Hume, Voltaire, *Kant,* Hegel, *Schopenhauer, Emerson, Spencer, Nietzsche, James,* Bergson, Santayana		

When one examines this matter objectively, using data discussed to some extent previously, it becomes apparent that mentally ill persons are actually greatly overrepresented in the ranks of great contibutors. Only 5 percent of men of genius should develop mental disease if no relationship existed between creativity and psychosis, but the actual rates are much higher. Seeing such scientific giants as Isaac Newton or Michael Faraday on the psychotic side of the ledger certainly makes one wonder about the burden concept of psychosis. The previously mentioned studies of famous artists and authors lead to a similar impression. Systematic investigations have made it clear that the psychosis-prone have indeed contributed to human progress far in excess of their proportion in the population. If at least one-fourth of men of genius develops overt psychosis, it seems probable that they all must carry the schizophrenia gene, since penetrance of psychosis is always incomplete.

In the comparative assessment of persons who engaged in scientific endeavors, it is usually more difficult to identify a specific discovery with an individual than is the case in arts or literature. A famous painting is credited to one artist, and so is a widely-read novel. Scientists, on the other hand, must always build upon a preexisting foundation, each successful individual usually contributing just one additional aspect, which may enhance an overall design. Lee De Forest was lucky when his improvement of the vacuum tube made it a highly useful instrument almost overnight, paving the way for world-wide radio transmission, but he could not claim to have actually invented the original vacuum tube. It thus becomes much more difficult to select scientists rather than artists and be certain one is picking the top contributors. Although contemporary scholars are often listed as eminent on the basis of the positions they occupy or prizes they are awarded, historical recognition follows different paths, often giving credit to those who were rejected by their peers.

In the ranks of scientists, no one is likely to debate that Isaac Newton, whose psychotic symptoms were so definitive that he was unquestionable schizophrenic, towered over other men in theoretical physics. One thus finds the statement that "Newton is beyond dispute the greatest scientist who ever lived, the only one of whom it can be said: had he not lived, the course of science might have been radically altered." Such facts alone should put to rest the notion that schizophrenics have been a burden on society. The greatest physicist of recent times, Albert Einstein, was also a strange man, said to have suffered a nervous breakdown in his youth, and he had a schizophrenic son.

Having mentioned Newton, Faraday, and Einstein, one only needs to add Robert Mayer, who discovered the law of conservation of energy, to indicate that scientists who became mentally ill have indeed left their marks in the field of physics. A systematic study of famous men in the development of mathematics substantiates that here, too, psychosis is common, frequently taking the form of mental depression. Similar surveys in other fields of creativity, including philosophy, music, literature, or art (covered partly in Table 18-1), soon convince one that there is a pattern to these findings. There simply exists no area of human endeavor in which the chief contributors have not shown a high rate of mental illness.

Since the present study is in the field of genetics, it seems appropriate to mention that Gregor Mendel, who was a monk, actually did not start out as a man of religion, but was forced to choose that station in life because of repeated nervous breakdowns, which blocked other preferred paths. Even Charles Darwin was a sick man, several authorities having concluded that he suffered from manic-depressive illness.

It is well known that institutionalized mental patients often engage in creative behavior in fields such as writing, poetry, music, or painting. The tendency is, however, to view such activities with criticism, considering them idle, useless, or even grandiose and delusional. But those who are willing to see can find plenty of support for the thesis that creativity carries with it the seeds of mental illness. Those who prefer not to see will remain blind.

Social issues concerning mental patients

In view of the convincing evidence that carriers of the proposed schizophrenia gene may be serving the human race as its chief source of creativity, society should now recognize that mental patients and their relatives have indeed made important contributions. Without their genes man might still live in caves. The families of schizophrenics thus have every reason to hold their heads high and aggressively oppose the unsupported accusation that they are somehow inferior. There is no justification for such families being stigmatized, and they should proudly make a claim to their rightful place in society. If the world is willing to take advantage of the contributions of psychosis-prone persons, there is no need to heap abuse on their relatives.

It also seems a shame that psychotic patients must be mistreated and degraded. If they are in fact paying the price for man's cultural advance, their suffering should be a concern of all. At present they are herded into institutions, where through legal measures they are often forced to undergo unproven "therapies." This has sometimes involved actual physical assault on them, even enforced mutilating operations. If one recognizes the value of the schizophrenia gene to mankind as a whole, it seems that there should be legal protection for mental patients against any involuntary "treatments" which cannot be scientifically validated.

Only one form of therapy has been thoroughly established to be of assistance to the mentally ill, namely the now widely practiced pharmacotherapy with antipsychotic medications. This form of treatment also fits the new scientific concepts. But the world is full of zealots who claim that they can cure disease by simplistic or punitive methods, and unfortunately society finds it expedient to unleash them on mental patients. Much of this activity belongs in the area of quackery and is carried out by nonmedical personnel, who are making a living for themselves at the expense of mental patients, while refusing to accept evidence which contradicts their manipulations.

According to the basic formulation presented in this book, there is a price for being productive, such persons having an increased risk of abnormalities in their families. This is a burden which mankind must bear, but there seems little reason for punishing those who became the victims of the unfortunate and unavoidable flaws in man's biologic structure.

It has not been possible to determine as yet whether pharmacologic treatment aimed at psychosis interferes with creative productions. When one is dealing with a seriously ill patient who shows no signs of creative inclinations, little can be lost by giving him the benefit of available therapy. It also seems clear that the majority of severly psychotic patients is devoid of creative tendencies. However, some creative persons have continued to be productive after the onset of psychosis. Abraham Lincoln suffered his first psychotic episode as a teenager and still went on to make important contributions. Gregor Mendel was also in his teens when he first became ill. Van Gogh produced his best paintings after he became psychotic. One cannot assume that a psychotic break ends all likelihood of creative work.

It seems, however, probable that the availability of effective treatment may at times indeed reduce the chances of creative production. In particular, persons who feel miserable and depressed, as Darwin did, are inclined to seek and accept therapy that lessens their suffering, and when the excess brain stimulation is turned off, the creativity is likely to be curtailed. To counteract losses that are thus likely to result from treatment, society needs to recognize those who possess a creative impulse and encourage their productions while they are still fit, instead of hindering them, as so often has been the case. There may well be social benefits in better care for the mentally ill, assuring them of their worth, instead of degrading and shaming them as currently is the vogue. All of society will be enriched if creativity is encouraged.

19
Physiologic Regulation of Brain Arousal

The chemical intricacies of the functioning of the human brain have been difficult to unravel, because of the complex anatomy and physiology of the central nervous system, coupled with its relative inaccessibility in the living state for structural or chemical studies. In the evolutionary process of the vertebrate brain, starting with a primitive organism and gradually building toward greater complexity, nature has never had the opportunity to scrap the old system and start anew with a less cumbersome design.

Localization of brain functions

The relatively undeveloped brain of ancestral forms had to possess all the basic physiologic modalities to enable the organism to respond to stimuli, avoid danger, search for food, and carry on reproduction, while at the same time maintaining proper balance or homeostasis of essential body functions. Later forms, with increased brain powers, had to build onto the preexisting system, adapting newly developed structures to the machinery already in operation. In the human organism, which possesses a brain with increased voluntary control, greater reasoning powers, and ability for verbal communication, the cerebral structures still operate through primitive areas, which remain an integral part of the central nervous system. It is therefore understandable that many of the neural pathways are quite indirect, that overlapping functions occur, and that some areas must be properly inhibited at times to keep them from interfering with the newer systems. Neuroanatomy thus is a highly complex field, far from being thoroughly understood, even with the availability of present-day tools for exploration.

One source of useful information about the detailed structure of the brain has been the frequent occurrence of localized abnormalities. Sometimes these are congenital, but more often they are caused by accidents or disease. Blocking of a blood vessel may result in loss of brain substance in that area, giving neurologists an opportunity to study the manifestations of the resulting defect. When the patient dies, it becomes possible to dissect the brain and determine the exact site of the lesion, and thus the normal function of that region can be mapped out. Brain tumors serve the same purpose. In this manner rather precise information has gradually accumulated about the operation of the different areas and the interrelationships of the various structures.

Neuroanatomists have also been able to establish many of the important nerve fiber connections through microscopic studies of sectioned and properly stained brain tissues. Use of experimental animals has further contributed to the overall knowledge, making possible the artificial induction of lesions, the introduction of appropriate chemicals, and the preparation of living specimens, which can be subjected to experimental procedures. The combination of all these approaches has led to a great deal of anatomical knowledge.

Chemical transmitters in the brain

In recent years another field of neurology has assumed increased importance, because of advances in methodology. This is the entire discipline of neurochemistry and neuropharmacology. The scientific activity in this area is at present very exciting and is leading to great advances in the total knowledge about brain function. Entirely new physiologic concepts have evolved in the last few years as a result of these developments.

For a long time acetylcholine was the best known neurohormone, and this chemical was thought to be largely responsible for nerve transmission within the brain. Acetylcholine is a relatively

Physiologic Regulation of Brain Arousal

simple substance, which is synthesized inside the cholinergic nerve cell and released at a nerve ending, when the electric impulse has traveled along the fiber to its termination, where that cell needs to transmit the message to a neighboring neuron. When the second cell has responded to the chemically triggered stimulation, initiating an impulse to travel along its fibers, the remaining acetylcholine is inactivated by an enzyme which splits it into its two major components, and thus the stimulation ceases. Unfortunately, acetylcholine has not lent itself well to microscopic studies to demonstrate its localization, as methods to detect it in the tissue are rather unsatisfactory.

Acetylcholine is generally associated with the voluntary nervous system and with the parasympathetic (cholinergic) aspect of the autonomic nervous system. It has also been known for a long time that the opposing neurohormone, norepinephrine (noradrenalin), usually thought of in connection with the sympathetic (adrenergic) part of the autonomic system, can be demonstrated to be present in the brain, although in a smaller quantity. For some time it was assumed that this latter substance was in the brain mainly to regulate the state of contraction of the blood vessels, since the sympathetic system was known to serve that function. In recent years it has, however, become clear that norepinephrine and its immediate precursor, dopamine, both have important actions of their own within the brain substance. Previously it had been assumed that dopamine existed only as a source that could be converted readily into norepinephrine, when the latter was called for to regulate the tone of blood vessels.

The areas of the brain which utilize dopamine or norepinephrine as their transmitters are actually the more primitive regions, shared by all vertebrates, including fishes, amphibians, reptiles, birds, and mammals. As evolution has resulted in progressive enlargement of the cerebral hemispheres, which in higher forms carry out the functions of more precise control over voluntary muscles, association of information, and application of reasoning, the older areas have become buried inside the center of the huge newly developed brain mass, but they still carry out their vital functions, including the regulation of body homeostasis, preparation of the organism for self-defense, and control of the state of arousal.

Other chemical transmitters, thought to serve important functions in the brain, are also being investigated, but their importance in relation to the subject under discussion here is not understood.

Adrenergic nerve transmitters

The catecholamine neurohormones, a term often applied to both norepinephrine and dopamine, as well as epinephrine (adrenalin), contain a highly reactive ring structure, which lends itself well to detection and measurement in microscopic preparations using fluorescent histochemical techniques. This has made it possible to map out the brain areas which contain these substances and trace their connections with other regions. It has thus become apparent that important brain areas exist which function through such transmitters. Some of these areas involve body movements and coordination, but others, of prime interest in connection with intelligence, are concerned with states of arousal and tone of awareness, obviously of central importance in relation to alertness, responsiveness, or creativity.

One interesting difference in the operation of catecholamine neurohormones, as contrasted to acetylcholine, is that once the transmission of the nerve impulse from one cell to its neighbor has occurred, the stimulation is terminated mainly by reuptake of the hormone into the cell, rather than by enzymatic splitting into inactive components. This reuptake process therefore becomes another vulnerable spot where something can go wrong. It should also be reiterated that quantitatively the areas served by acetylcholine are much larger in the human brain than those utilizing catecholamine transmitters, as they include the large cerebral hemispheres.

The rapidly expanding knowledge about the various neurohormones has opened up new frontiers for exploration and pharmacologic experimentation connected with brain physiology. These developments happened to occur at about the same time as people became more interested in the availability of various centrally stimulant drugs, such as hallucinogenic agents. These two areas have overlapping regions, which have served to elucidate certain important aspects of brain activity.

It is not the intent of the present work to attempt to cover the available knowledge about psychedelic drugs, as many books have been written on that subject. However, it is of great interest that the evidence indicates that such drugs may act through the catecholamine systems described above, the current thinking being that dopamine is the chemical most intimately involved.

There is also much evidence that overactivity of the dopamine system may be related to proneness to psychosis. For one thing, one can hardly remain unimpressed with the well confirmed observation that amphetamines or pep pills, which are known to activate the dopamine mechanism, are capable of reproducing all the signs of a schizophrenic psychosis. Some individuals, who have consumed excessive amounts of amphetamines over a period of time, enter mental hospitals with symptoms that often are indistinguishable from paranoid schizophrenia. This condition has been referred to as amphetamine psychosis, and it can be duplicated by excessive use of cocaine, another centrally stimulant drug. Since it is now fairly well established from other studies that these substances exert their activity through enhancement of dopamine action, the findings lend strength to the dopamine hypothesis of schizophrenia. Still further support comes from the observation that psychosis often develops in patients treated for parkinsonism with L-dopa (L-dihydroxyphenylalanine), a precursor converted in the brain to dopamine.

The field of comparative neuroanatomy gives additional leads in understanding the evolution and present operation of the adrenergic areas of the brain. In the developmental process, the central nervous system arises from the ectoderm, the embryonic layer which also forms the skin. The ectodermal cells contain melanin, a pigment which can range in color from black through brown to lighter shades, as its concentration is varied. In many primitive organisms there exists a mechanism for displaying variable amounts of the pigment, depending on external circumstances and need for protective camouflage. The early types of nervous systems presumably arose from such structures, which already had the capacity to respond to external stimuli.

Melanin is formed from L-dopa, and thus there has always existed in ectodermal cells a precursor which by a small chemical change can be transformed into dopamine. This is shown in Figure 19-1, which illustrates the metabolic interrelationships of the catecholamine hormones. As the nervous system evolved, it is not surprising that dopamine thus may have been the original adrenergic neurohormone, therefore being found in the most primitive regions, even in the brains of higher forms. As additional systems were later added, one can conjecture that conversion of dopamine to norepinephrine made it possible to control new functions, utilizing the latter hormone, without interfering with actions already dependent on dopamine. A further step converts norepinephrine to epinephrine (adrenalin), which presumably was the last of these neurohormones to evolve.

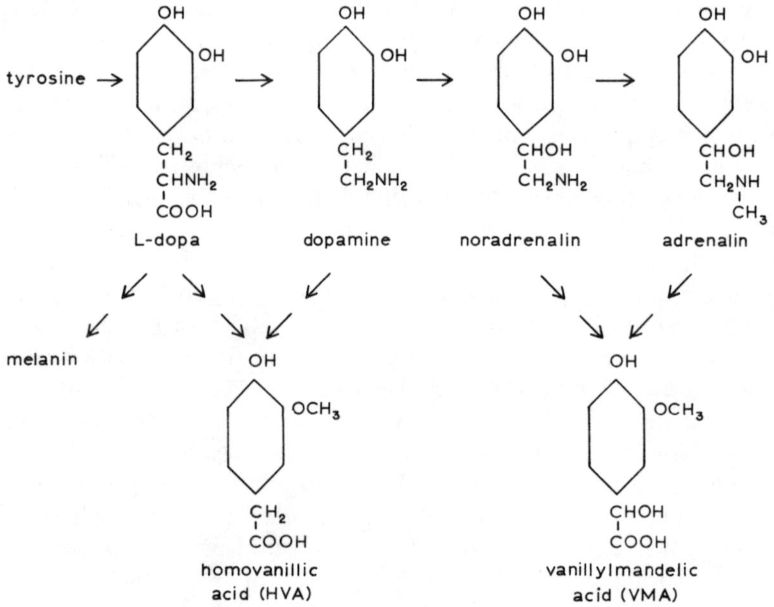

Figure 19-1 Origin and fate of catecholamine neurohormones. HVA and VMA are excretion products.

Placing these chemicals in an evolutionary context enables one to explain why dopamine carries such a unique and central role, in spite of the fact that other hormones are quantitatively more abundant and qualitatively more related to higher brain functions. This hierarchy also suggests the possibility of treatment for dopamine disorders without interfering excessively with other brain mechanisms. It indeed seems fortunate that a distinct chemical appears to be involved in such a fundamental capacity, which, therefore, can be studied and dealt with as a somewhat isolated system.

Dopamine antagonists

Antipsychotic medications, including the phenothiazine tranquilizers, apparently act on the sympathetic nervous system in exactly the opposite fashion of amphetamines, and are therefore classified as adrenergic blocking agents. Patients suffering from amphetamine psychosis can be effectively treated with phenothiazines, although often this is unnecessary, since simple removal of the offending drugs soon leads to subsidence of the symptoms. In a way the term "major tranquilizers" is unfortunate as a designation of the phenothiazines and related medications, although it is true that in the normal person, or in the psychosis-prone individual during his healthy intervals, these medications indeed tone down the activity of the brain, inducing a state of greater tranquility. In this role they can be utilized for chemical restraint, when this effect is desired. However, when these pharmaceuticals are used to treat schizophrenia, they serve in an antipsychotic capacity, although this is thought to be achieved through the same chemical mechanism, presumably by reducing the activity of dopamine. The sick individual may be withdrawn and inactive or depressed, and bringing him to a more normal physiologic state may result in increased mobility, so that tranquilization is not always the observed effect.

The precise mode of action of antipsychotic medications is still unknown, but considerable knowledge has been accumulated. The most widely accepted hypothesis is that these chemicals attach themselves to the receptor sites of the nerve cell intended to receive the stimulus as dopamine is released, thus blocking the neurohormone from doing its work. The evidence for this is mostly indirect, one observation being that in the early phases of treatment with phenothiazines the amount of the excretion product (HVA) of dopamine is increased. This has been interpreted as a sign that in its attempt to pass on its message the first cell continues to release additional hormone. Another sign, seen as evidence of receptor blockade, is the recording by means of microelectrodes of enhanced neuronal firing activity in the dopaminergic nerve cells under the influence of phenothiazines. However, a recent report by Seeman and Lee suggests that rather than interfering with the receptor sites, the phenothiazines may perhaps reduce the amount of dopamine released from the first cell. If this is confirmed, one would have to assume that the excess HVA results from diversion of the synthesis within the blocked cell, and that the increased firing rate also is caused by that blockage. One may also wonder whether schizophrenia perhaps results from an abnormality in the reuptake mechanism for dopamine. Such an abnormality would lead to increased amounts of dopamine in the interneuronal cleft, which could be alleviated by a reduction of dopamine release or curtailment of the activity of the hormone.

It is obvious that important developments in the understanding of normal brain physiology are taking place in this exciting field. Amphetamines and many psychedelic drugs, according to current knowledge, increase dopamine activity, thus "turning on and tuning in" the brain. Many of the proponents of psychedelic drug use have claimed that increased creative activity is one of the effects. On the other side of the scale, the major tranquilizers tone down the state of arousal and reduce creativity.

It is interesting that most of the potent antipsychotic drugs have related chemical structures. They all contain a ring arrangement, either the phenothiazine nucleus, similarly complex derivatives, or even simpler substituted six-membered rings, the latter having been viewed as half of a phenothiazine nucleus. This part is attached to a side chain, which generally consists of three carbon atoms terminating with a nitrogen. The latter then has simple substituents or branches leading away from it. Some of the chemicals contain complicated attachments at the distal end, which are not necessary for the antipsychotic activity, although they increase potency.

Except for recent derivatives, which have led to some changes in the three-carbon side chain, probably without disturbing the overall architectural relationships, all the active antipsychotic organic chemicals have the same basic structure. Presumably this type of chemical can somehow interfere with the release or activity of dopamine, and it has been demonstrated with radioactive tracer techniques that such chemicals indeed compete for certain receptor sites with the latter. It has also been shown that the overall stereochemical arrangements are sufficiently similar between the adrenergic neurohormones and the antipsychotic drugs to postulate that one could substitute for the other on an enzyme or receptor site, and in this manner the medication can block the action of the hormone. Structural and physiologic studies of the drugs thus give still further support to the dopamine hypothesis.

Social impact of antipsychotic medications

The discovery of the phenothiazine tranquilizers more than

twenty-five years ago ranks with the greatest medical achievements of all times. Mental illness affects so many people that any effective treatment in this area benefits a very large number. Until the discovery of the phenothiazines, there was no effective therapy available for mental disease, and the number of mental patients in hospitals had risen each year, essentially in proportion to the growth of the population.

The psychiatric profession was, however, unprepared for a truly medical treatment of mental illness. The standard psychologic theories, which in reality amounted to little more than a rehash of old beliefs that had existed for centuries, assumed that there was no organic basis for schizophrenia, and derangements of the mind were conjectured to occur through improper developmental input. Such hypotheses obviously did not predict an effective pharmacologic treatment.

Interestingly, a report on the response of psychotic patients to an adrenergic blocking agent was published several years before the discovery of the phenothiazines, but no one paid attention to it. Rockwell found dibenamine, one of the early medications in this class, to be effective in controlling hallucinations and other symptoms in patients who had been sick for many years. Even authorities in pharmacology dismissed this observation as no doubt caused by secondary effects, and it was not followed further.

When the phenothiazines had become available, psychiatrists still were not about to accept that mental illness could be treated with drugs. First the effectiveness was questioned, and the doctors of the mind, who always in the past had accepted any proposal for therapy rather uncritically, insisted on blind tests, followed by the double blind design where neither the patient nor the doctor knows what is being prescribed. Even when such rigorous comparisons had supported the effectiveness of pharmacotherapy, the preferred explanation remained that the medication only served to calm the patient, making him more amenable to psychotherapy, which still was seen as the real treatment. Although the scientific evidence indicates clearly that patients do just as well with or without psychotherapy, as long as they get their medications, this formulation remains the prevailing one in psychiatric circles. After all, what would clinical psychologists do for a living if it was recognized that schizophrenic patients only need their medicines and general support? However, if one accepts that they only suffer from a defect in the regulation of arousal, while the brain itself is intact, psychotherapy indeed appears to be the wrong prescription.

Many variants of the phenothiazines are now available, and the evidence indicates that they are mostly interchangable in their antipsychotic effectiveness. Other major tranquilizers, which do not contain the phenothiazine nucleus, appear to have the same basic action. The inorganic ion lithium, which now is often used to treat manic-depressive illness, is also thought to reduce the production of dopamine in the brain. Most mental patients are at present treated with these medications, even those doctors who still cling to psychodynamic theories finding it expedient to administer chemical treatment to their patients, knowing that otherwise they are unlikely to improve. Unfortunately, however, the patients in this setting are given an explanation of the purpose of the medication which is not conducive to acceptance. If it is presented to them as a tool intended only to control their undesirable behavior, rejection is invited.

The question whether a person who develops psychosis should from then on take antipsychotic medications has not been settled. Certainly there are people who suffer a psychotic break and have no further difficulty, and there is no reason why they should continue to take drugs indefinitely. Others have episodes of illness and then are free of symptoms for long periods, perhaps needing no treatment during such intervals. Still others suffer frequent breaks if they take no medication. Many manic-depressives have been found to have fewer episodes of illness if they take lithium continuously. On the other end of this spectrum is the large number of mental patients who frequently must return to hospitals for involuntary treatment, because they or their therapists have discontinued all pharmacologic treatment. This is actually the main cause of rehospitalization for mental patients, and the responsibility for their suffering rests with those who refuse to recognize the need for continued medication. Some patients even commit serious crimes while deprived of treatment, and they may end up in prison for failures which could be blamed on their therapists.

Even after the antipsychotic medications had become available, proponents of the older theories continued to promote vigorously opposing concepts of treatment, such as community psychiatry or crisis intervention. Legislative approaches have been used to exclude physicians not trained in Freudian principles from the care of mental patients. But today the main therapy is actually pharmacologic, although claims are often made that the real treatment is still psychotherapy. It seems strange that those who claim that schizophrenia may be caused by double messages communicated within the family should choose to enmesh their patients in conflicting explanations of their own treatment procedures.

In view of the tremendous impact that antipsychotic drugs have had on the medical field, it is surprising that even after a lapse of a quarter-century there is no talk about giving a Nobel prize to the Frenchmen who discovered the phenothiazines. This is unquestionably one of the most important contributions of all times to medical science, and it should be so recognized. Strangely, there is only one contribution in psychiatry that has earned this high award, namely the development of prefrontal lobotomy. This is actually a mutilating brain operation, so destructive that the worst patients in mental hospitals, for whom nothing effective can be done, are those who many years ago were subjected to this procedure. Psychologists supported the claim that parts of the frontal lobe of the brain could be dispensed with without significant intellectual loss, but anyone can, without tests, see for himself how devastating the effect is in reality. It says something about administrators that they singled out this treatment for a Nobel prize.

Despite all the debates in the psychiatric field, one thing is certain: Antipsychotic medications are here to stay, and they have radically altered the outlook for mental patients. This entire field has also contributed in great measure to increased knowledge about the normal physiology of the brain, including basic factors involved in intellectual activity. Rational use of the antipsychotic medications is dependent on an understanding of genetic and physiologic mechanisms. Progress toward scientific explanations of creative behavior is intertwined with developments in the pharmacologic treatment of psychosis.

SECTION IV
Personality Genetics and Social Planning

20
Inherited Personality and Creative Aptitude

In the first section of this book it was pointed out that a discrepancy appears to exist between the rate at which genetic factors normally undergo a change and the estimated rapidity of the evolution of human mentality. To account for the unusually quick adaptive changes exhibited by prehistoric man, it was postulated that special mechanisms must have been utilized, perhaps mainly those involving heterozygote advantage or hybrid vigor. Progress in regard to improved mentation may then have been dependent on genetic systems which lead to concomitant disorders, the latter showing up most clearly in homozygous individuals, but also in a fraction of heterozygous carriers. Although hybrid vigor is a general principle, available to all higher creatures, it can only lead to an improvement in a fraction of the population, being therefore of benefit mainly to social organisms, its value further enhanced by communication skills.

If the proposed mechanisms have in fact been largely responsible for the development of mankind's exceptional brain potentials, there should also exist in all human populations a totally "normal" man, free of the otherwise abundant defective genes, but relatively limited in mental powers. Thus one expects to see a greater degree of diversity in human populations than in most organisms. Persons who carry a gene for myopia, for schizophrenia, or for diabetes, and in particular those who possess favorable mutant genes of more than one kind, should often exhibit increased cerebral activity, and the whole human community then benefits from their enhanced abilities.

The findings described in this book appear to give substantial support to this hypothesis. The way has been paved for further investigations of specific genetic factors involved in personality development. In a sense the more able person can be seen as abnormal, being burdened with nonfunctional mutant genes. Superior intelligence may in this context be viewed as somewhat analogous to genetic disorders, and its inheritance can consequently be studied by the methods usually applied to human diseases.

Personality effects related to specific genes

If several different mutant genes have established themselves at increased frequencies because of favorable effects on brain action, it is likely that each has a specific type of influence on behavior, and one then expects to find different personalities associated with the individual factors. One kind of gene may increase abilities in logical thinking, another in memory, still another in perseverance.

Perhaps in the future it will be possible to demonstrate a concentration of certain genes in selected groups of people, such as in mathematicians and scientists, in political leaders, or in members of the clergy. The tools are now available to search for such differences.

Several authors have claimed that myopic persons tend to be excessively introverted, but systematic investigations in the area are meager. Limited results of psychologic studies have indeed suggested profound personality differences between myopes and nonmyopes, but the data are too scanty to permit any definite conclusions. One study indicated that on vocational tests myopes showed greater academic tendencies, while nonmyopes in this group were more inclined toward business interests. Additional data are needed on personality traits of both myopes and myopia carriers, contrasting these with findings on individuals unlikely to possess the myopia gene.

Descriptions exist in the literature of personality traits that possibly can be attributed to the postulated alcholism gene. There is an impression that many alcoholics are friendly and sociable, but others are sociopaths or criminals. Perhaps a tendency to a high activity level is the common characteristic.

Studies done many years ago by Malzberg on mental hospital populations in New York indicated that people of Irish origin were excessively prone to alcoholism. This can fit the popular impression that the Irish are characterized by certain personality traits. Perhaps their friendly and outgoing manner is somehow connected to the alcoholism gene. But there may also be some truth in the notion that overaggressive and self-confident individuals, who force themselves on the public, utterly convinced that they know what is best despite an apparent pitiful lack of judgment, are similarly under the influence of this factor. Some leaders with such obnoxious qualities have in fact had drinking problems.

A link of alcoholism with hyperactive childhood behavior and adult sociopathy has been recognized for some time, this association having shown up in many independent investigations. Various other negative traits have been described, but a bias may have existed in populations studied in institutional settings. Further systematic surveys are needed of the traits prevalent in relatives of alcoholics, emphasizing the positive aspects. Assertiveness, leadership, and communication skills may well characterize such groups. Mankind needs effective leaders, and perhaps the alcoholism gene is a major source of their basic qualities.

A corollary of the genetic hypothesis of schizophrenia is that in each instance at least one of the parents of a psychotic patient must be a carrier of the proposed gene, whether or not that parent exhibits any signs of illness. One may thus expect to be able to find some associated personality effects. It is now well documented that psychotic patients as well as their siblings are more likely than the overall population to be born in the early months of the year, and it is easy to imagine that here is one sign of a variation in behavior exhibited by nonpsychotic gene carriers. Presumably, then, the carrier parents show increased sexual drive during the early summer months, children conceived during that period being born in the first part of the following year.

Other personality effects have been demonstrated by test procedures. Psychologists report that a significant proportion of all people, in the order of one-sixth, exhibits characteristics on the MMPI test which could be interpreted as indicative of psychotic tendencies, although no illness is apparent in most such individuals. Psychotic patients and their relatives show still higher rates of deviance. Abnormalities are also said to be increased in populations of college students, although until this time there has been no explanation of that finding.

Corresponding observations have been described in connection with modified versions of the Lovibond object sorting test, which requires the person being evaluated to arrange certain commonly known articles into groups and then explain the reasons for his choices. Schizophrenic patients, as well as many of their nonpsychotic relatives, have been found to react differently than the average person, giving responses considered to some degree "irrelevant." At one time this was seen as evidence of a thought disorder, but when it was also discovered in school populations that students who exhibited "disordered" thinking were often high-ranking members of their classes, it was decided that a better description would leave out the value judgment. One author has suggested the term "allusive" thinking for such tendencies, still implying that the desired answer is alluded to rather than being given directly. This evaluation is, again, obviously a matter of opinion. The occurrence of deviant modes of thinking in psychologic test responses of persons who are likely to carry the schizophrenia gene has in recent years been confirmed by several investigators, although there is considerable disagreement about the meaning of this observation.

Despite the documented relation between psychosis and creativity, there has for a long time prevailed an inclination to stamp a label of inferiority on schizophrenics and their families. Thus they are accused of giving double messages, distorting meanings, or creating conflicts. Even professionals paid to care for mental patients join others in such unwarranted discriminatory attitudes, which stem from no validated findings, but rather from a human tendency to degrade or slander potential competitors, thus enhancing the critics' own self-image at the competitors' expense. Although there appears to remain little doubt that carriers of the schizophrenia gene indeed differ in thought patterns from noncarriers, no scientific support exists for the view that the deviation is toward inferior thinking. In fact, it may well be in the opposite direction. If one defines "normality" as the characteristics exhibited by the majority and equates the term with "superiority," any deviation is by definition abnormal and undesirable. However, if a minority with a different thought pattern can be established scientifically to reason more precisely, despite that way being deviant, it seems unwise to condemn their mode of thinking as automatically somehow inferior. Scholars should not forget that a schizophrenic explained the nature of gravity, which had puzzled "normal" people for centuries.

It does not seem unlikely that specific tests will become available in the future, which will be capable of detecting carriers of the schizophrenia or alcoholic genes early in life. One report has described how individuals likely to be schizophrenia carriers react excessively to hallucinogenic agents such as LSD. Other reports have claimed a variation in their brain waves. A combination of drug and sleep research, using refined electroencephalography, may eventually reveal differences between carriers and noncarriers of brain related mutant genes. Then there is the possibility of finding specific patterns of chemicals in the body fluids of heterozygous carriers. Hopefully, when such possibilities materialize, they can be used constructively, and not merely for the purpose of identifying persons to be labeled as abnormal. A capacity for creativity or leadership is at least a possibility in carriers of certain genes, revealing their potential strengths, aside from vulnerability to disease. The possibilites along these lines are obviously promising.

It has been claimed that psychosis-prone persons are likely to team up as marriage partners, because of the personality effects of the genes. However, systematic studies have shown little increase in psychosis in the spouses of mental patients. A recent report suggests that schizophrenic patients may often have sociopathic mates, but once more no increase was found in psychosis. There is also an impression that alcoholics tend to marry each other.

Interaction of different personality genes

Acceptance of the view that a truly creative inclination depends on the possession of a schizophrenia gene, the personality then being further modulated by other genes and by environmental influences, leads one into new areas that would not be likely to be explored in the absence of this conceptualization. A related area of inquiry concerns leadership qualities.

One naturally wonders why those who have proven themselves capable of developing novel ideas, on which human progress appears to depend, are seldom chosen as leaders. In fact, there is a popular image of such persons being unsuitable for leadership. Of course one of the factors involved here is that the public must support a leader, and a man who does not fit the proper image and consequently fails to inspire the confidence of the masses is not likely to become an effective leader. The characteristics seen in creative individuals are usually not those pictured by the public as the qualities of the man in charge. While the creative person can thus be a good thinker, he may not in most cases be destined for leadership. This is not the same as saying that those who end up being chosen as leaders necessarily show a high degree of wisdom once they have acquired the power, but rather that they present the image that the public for some reason favors.

Creative intelligence is the crucial ingredient on which cultural progress depends, so that recognition and support of creative persons obviously is in the interest of all of society. Whether any agreement on methods to ascertain creative intelligence is likely in the near future is difficult to predict. Facilities such as the Institute of Personality Assessment and Research at Berkeley were set up to develop new knowledge in this area, mainly because the military establishment needed methods to classify people, but so far there is no agreement on the success of that operation. Creative minds are frequently only recognized retrospectively, after they have accomplished their work without any outside assistance.

Since several specific genetic factors have been proposed in this book, each serving a fundamental role in personality development, the question naturally arises how these factors may interact with each other in the molding of the total personality. There has always been a tendency to think of people as falling into recognizable categories, such as leader types, business types, or criminal types, and perhaps some analysis is now possible of a biologic basis for such divisions.

Although no information has been dealt with here which ties a diabetes gene into mental development, there are indications that such a relationship may exist. Diabetics and their relatives are known to have a tendency to be macrosomic and overweight. While the explanations are usually developed in reverse, overweight making for a risk of diabetes, the possibility that a diabetes gene may as its principal effect cause increased growth and a gain in weight, for example through overproduction of steroid hormones or excessive release of growth hormone, has never been explored adequately. The effect on the pancreas may then be a secondary phenomenon, produced by strain on that organ resulting from an altered metabolic pattern. If the proposed diabetes gene influences the somatic constitution, it may also have an effect on mental traits. It is not uncommon for diabetics to develop mental depressions, although here again the preferred explanation has been that being a diabetic would make one feel sorry for oneself.

In comparisons of different populations it has been consistently reported that individuals of Jewish origin have an increased risk of diabetes, while they have a decreased rate of alcoholism. One can thus conjecture that the diabetes gene is somehow related to personality traits thought to exist in Jewish people. A study done in France has led to the conclusion that large individuals achieve slightly elevated scores on IQ tests. Books have been written to document how successful persons tend to be above average in size, the favored interpretation being that this is evidence of discrimination against those of smaller stature. If one postulates, however, that diabetes carriers are characterized by larger body size and more active brains, these puzzling facts can fit together. Perhaps some diabetes carriers become effective in public relations, being above average in extraversion.

Among psychotic patients, those with a manic or depressive form of illness have a tendency to be robust, while individuals with a typical schizophrenic expression are more likely to be slender. One can thus wonder whether manic or depressive patients may not carry a diabetes gene, or a similar factor, in addition to the gene for psychosis. It is actually not uncommon to see overt diabetes in patients suffering from affective psychosis. A dichotomy may therefore be developed, patients who are carriers of a schizophrenia gene either having or lacking a diabetes gene and thus falling into the rounded extraverted or trim introverted categories. Adding the alcohol gene to this system, and assuming that it produces hyperactive behavior, each group would be further subdivided into active versus passive types. The four resulting personality subgroups, related to psychotic tendency, are listed in the top portion of Table

20-1, carriers of a schizophrenia gene being referred to as hyperphrenic.

Table 20-1

Hypothetical scheme for classification of human types in terms of genes for alcoholism, schizophrenia, and diabetes (Aa=alcoholic carrier, Ss=schizophrenic carrier, Dd=diabetes carrier)

Genotype	Personality designation	Popular description
$AaSsDd$	Active hyperphrenic extravert	Manic depressive
$AASsDd$	Passive hyperphrenic extravert	Unipolar depressive
$AaSsDD$	Active hyperphrenic introvert	Paranoid
$AASsDD$	Passive hyperphrenic introvert	Catatonic
$AaSSDd$	Active normophrenic extravert	Leader
$AASSDd$	Passive normophrenic extravert	Professional (?)
$AaSSDD$	Active normophrenic introvert	Sociopath
$AASSDD$	Passive normophrenic introvert	Normal person

Although this division is presented here mainly to illustrate one type of approach that may be useful in this field, while meaningful data for its support are at this time largely missing, it may be pointed out that the resultant system is not inconsistent with available knowledge. The psychoses are usually divided into the affective illnesses on the one hand, including manic and depressive syndromes, and the schizophrenic disorders on the other hand, the latter often grouped into four subtypes. Since the simple type of schizophrenia is rarely seen, and investigators now agree that hebephrenia is in reality an advanced stage of severe or untreated psychosis, the basic subgroups of schizophrenia can be viewed as being really just two, either paranoid and active or catatonic and withdrawn.

It also remains to be demonstrated whether it is correct that in psychotic patients the alcoholism gene tends to produce either manic or paranoid behavior, while its absence leads to unipolar depression or catatonia. Psychiatrists know, however, that manic patients often drink, while there is no general increase in alcoholism for all psychotic patients. Recently it has been reported that the enzyme monoamine oxidase is found in reduced quantity in blood platelets of alcoholic as well as manic depressive or paranoid patients, this deviation being absent in other psychotic persons. If confirmed this finding would support the scheme in Table 20-1. Catatonic schizophrenia is for some reason more common in Orientals, and they are less inclined to consume alcohol.

If the proposed scheme has some validity, it also follows that the noncarriers, in relation to the schizophrenia gene, would fall into corresponding categories. The "normophrenic" group can thus be divided, on the one hand into the rounded and affable leader versus the less aggressive types, and on the other into the trim and tense sociopathic versus normal groups. These categories are also listed in Table 20-1. Perhaps carriers of the schizophrenia gene who do not develop psychosis resemble these latter persons in some ways, but presumably they have a greater potential for creative thinking. If one also includes in the analysis the schizophrenia homozygotes, who are rather rare and presumed to be chronically ill, they should similarly fall into four types, but such a division is at present of little importance.

Interestingly, the myopia gene cannot be fitted into the system proposed above, as there is no indication that myopes fall predominantly into certain basic personality categories. Myopia appears in all groups of psychotic patients, the nearsightedness rate in their ranks being the same as in others. Perhaps the myopia gene confers an intellectual advantage on anyone, without drastically changing his basic personality.

It should be reemphasized that the scheme in Table 20-1 is not presented as something proven to exist, but rather as a possibility that should be further explored. Adjustments of the hypothesis will no doubt be needed as new data accumulate, but an approach of this general nature may well be fruitful. Psychologists in particular should be interested in this kind of thinking. If the genetic basis of the subtypes of psychosis can be explained, this can serve as a framework for more general studies in personality genetics. It must also be kept in mind that additional genetic factors are likely to play a role in mental development.

In connection with proposals for genetically distinct human types, it may be mentioned that certain pharmacologic manipulations have suggested parallel possibilities. Everyone knows how alcohol can bring out behavioral traits which otherwise are not evident. In work with psychotic patients it has been observed that minor chemical adjustments can often produce dramatic personality effects. A British psychiatrist described several years ago how mentally ill women, who were leading a rather chaotic life despite treatment with phenothiazines, became more organized when they were also placed on birth-control pills. A more recent report, dealing with psychotic males given androgenic hormones, indicated that this brought out previously nonexistent paranoid ideations. These studies suggest that certain steroid hormones play an important role in personality characteristics. There are also reports of intellectual enhancement in girls exposed to androgenic steroids during intrauterine life as well as in children treated prenatally with the pregnancy hormone progesterone.

It has been reported that the drug physostigmine, which inhibits breakdown of acetylcholine in the brain, can terminate the euphoric behavior of manic patients. Observations of this type give strength to the view that simple biochemical differences are capable of expressing themselves as profound variations in attitude or behavior. It is thus not far-fetched to think that corresponding genetic differences can be important in the determination of personality traits.

Judgment and human intelligence

While no one denies the importance of adequate learning ability, and everyone is impressed with a mind capable of retaining and using a large amount of information, the essence of superior human intelligence is good judgment, even though there is limited agreement on the precise meaning of that term. Wisdom does not seem to be measured by IQ scores or scholastic tests. As an example, one can take professionals, most of whom have a high IQ and excellent learning ability as well as a tendency toward assertiveness. In terms of retention of facts and facility for utilizing information at the usual level of operation, limited differences are demonstrable between them. Still, other levels exist, where very significant variation is evident. Thus the kind of scientific reasoning elaborated in this book apparently can be appreciated only by a small fraction of mental health professionals. Since the factual material is not very complex, it is actually hard to imagine that ability to follow the arguments is in truth the limiting factor. More likely, it seems, the required intellectual curiosity is missing or, alternately, the flexibility needed to make a change from former modes of thinking. Ability in judgment ties in with these latter qualities. Emotional blocking can also interfere, but it would not be proper to suggest that experts in treatment of the mind could be so affected.

It is, on the other hand, difficult to conceive that someone possessing genuine creative intelligence could remain satisfied with superficial explanations of the past in the face of illuminating and challenging new information. But evidently those who possess good learning ability are not necessarily endowed with a creative component, and thus they may be unable to adapt to new ways, lacking true wisdom.

Just as there is no effective way at present of measuring creative capacity, no standardized methods are available to ascertain superior judgment. Whether wisdom is truly tied in with creativity cannot be scientifically decided until each basic quality can be defined more precisely. However, ability to arrive at a logical answer often seems linked to a knack for formulating the appropriate question, and it is not unlikely that such a quality is dependent on creative inclinations. There is thus some reason for believing that the best thinkers may be those possessed of creativity in addition to good learning qualities. Perhaps persons with a creative constitution can be good thinkers, even when little new production is manifest.

According to the computations presented earlier, one person in six carries a schizophrenia gene, and obviously such individuals cannot all be highly creative in the sense of changing the world order. Perhaps some are only good thinkers, showing their originality in more limited ways. However, historical evidence suggests that persons capable of supremely brilliant reasoning, like that exhibited by Newton and Einstein, must harbor the schizophrenia gene. Presumably such individuals carry additional important genes which remain unidentified.

21
Continued Integrity of Man's Genetic Heritage

The value of the investigative strategy formulated in Chapter 1, leading for the first time to an identification of specific genes involved in personality development, can no longer be disputed. It seems certain that further discoveries will be made along these lines, and this in turn may alter the prevailing philosophy toward genetic influences in human mentality.

Additional research into the exact nature of those genetic effects now becomes vitally important, and great strides may be expected in that area. If children are born with their basic mental endowments, rational efforts should perhaps be made to promote a better quality of life through genetic planning, instead of just accepting what comes and trying to make necessary amends. Such thinking has in the past been referred to as "eugenics," a term that now has a bad connotation, although for no valid reason.

Man's genetic legacy does not truly belong to the present generation; rather it should be seen as the property of those destined to live in the future. Currently-living people are the first to seriously interfere with nature's system of protecting man's genetic endowment, without having earned any license to meddle in that area. Under the existing circumstances it takes unselfish individuals to carry the banner of the future, as they are certain to be abused by unthinking critics who insist on unabridged freedom to consume or destroy whatever they desire. However, the rights of the unborn should somehow be protected.

Regulation of reproductive patterns

Early in this century, when the eugenics movement was new and rather popular, specific plans were implemented in several countries to attempt to rid mankind of traits considered detrimental. This approach, however, was premature, suffering from two principal weaknesses: First, the available knowledge was inadequate for genuinely valid conclusions, so that issues that on the surface seemed simple turned out to have other aspects, which needed to be taken into account. Secondly, this philosophy invited abuse, providing a tool for governmental agencies to justify politically-motivated attacks on those classified as undesirable.

The shortcomings of the eugenic approach soon became evident in these early attempts to implement policies that originally had seemed sound. For some time the opinion was, however, prevalent that families with defective genes should be sterilized in order to free mankind of scourges like mental illness, diabetes, or criminality. It is now apparent that such specific plans were often based on misconceptions, and in retrospect it seems that rather than eliminating the genes associated with frequent disorders, society should perhaps in some instances actively promote their propagation. However, there still exist deleterious genes that mankind is better off without.

A more recent example of a failure to adequately explore the basic biologic concepts before taking action is the current approach toward arterial hypertension. Afflicted persons may die relatively young of strokes or other complications, and the proper recommendation appears to be appropriate treatment to lower the blood pressure and ensure greater longevity. Little consideration has been given to the possibility that the factors leading to hypertension may actually be useful to mankind, and there indeed is some evidence in favor of such a view. Although in this instance no one is promoting sterilization of the affected persons, further studies are needed of hypertensive individuals and their contributions before society embarks on a course of controlling this condition. If treatment leads to a curtailment of productivity, lengthening of the life-span may be little compensation. Life is not measured in time alone, and hypertensives should not be labeled as biologically inferior until all the facts are established. This discussion obviously does not apply to the pathologic variety of hypertension which is immediately life threatening, but rather to the more generally encountered mild forms.

In spite of the existence of glaring weaknesses in man's total knowledge about his own biology, there is no justification for using them as an excuse to abandon the entire philosophy of eugenics. Some form of rational planning, even if it is far from perfect, is likely to be better than a drift with no direction. The latter course may in fact be disastrous to mankind's future.

At present many societies are suffering from upheaval of previous reproductive patterns, the changes having come about by accident rather than through social planning. Educated individuals are reported to be having few children, while lower social groups are taking over in greater numbers. If this were a natural course, which mankind had followed for some time, biologists would be inclined to move cautiously in recommending interference. But the present system is itself artificial, society having indirectly promoted the propagation of those who in the past found themselves unfit in the face of competition.

To establish a baseline against which one can assess the magnitude of recent changes in reproductive patterns, a study has been undertaken, using the Icelandic population as a model. Specifically, data were gathered to determine the relationship of the number of offspring to achievement, focusing on persons living approximately a century ago. There is no reason to believe that family compositions differed during this period from long-established conventions in Scandinavian populations, and Iceland has been an essentially class-free society.

Surviving males born in the interval 1851-1880 were randomly selected, and for each such male the total number of children alive after age fifteen years was ascertained. A determination was also made of the rate of listing in *Who's Who in Iceland* for the different groups of males. The results of this study are summarized in Table 21-1. It is clear that during the chosen period the more successful men were the ones who raised large families, and these findings are in agreement with older studies performed on Asian populations.

The data show one-third of males who lived into adulthood producing no offspring, and this segment is relatively undistinguished. Their failure in reproduction is made up by the socially most prominent group, which produces more than its share of surviving children. In this manner there appears to have been a constant displacement in past times of the less able by a more accomplished segment of society. The findings are consistent with the notion that some of the genes maintained at high frequencies by differential fertility and survival may indeed have an intelligence effect.

Table 21-1

Rates of listing in *Who's Who* in relation to the number of surviving offspring for Icelandic males born in the period 1851-1880

Number of offspring	Total males studied	Rate of listing in *Who's Who* %
0	443	5.2
1	128	8.6
2	146	11.6
3	136	14.0
4	123	15.4
5-9	328	18.0
10+	51	29.4

Recent statistics from the more populous Western countries indicate that the largest families may now be arising from parents of low IQ levels, thus illustrating the dimension of the current change. The impression in the United States is that there may in fact at present be a very significant shift toward increased reproduction by the least educated segments of society.

There is obviously something wrong with an artificially induced system which discourages the reproduction of educated people, because they can have children only by making a considerable sacrifice, while the less fit are handed unearned income to care for each child they bring into the world. In fact, with the present system in the United States the most effective way for a poorly fit young woman to make a living is to get herself pregnant and thus become automatically entitled to a paycheck, assured that it will increase in size as the process is repeated. No one recommends starving children, but perhaps society should have a hand in the overall planning to tip the scale toward desirable genes. A guaranteed freedom from hunger should not necessarily be construed as a license to unlimited propagation. Future generations will pay a heavy price if politicians now in charge are permitted to chart the course which is easiest for them to avoid controversial issues. While the current approach of dealing with the immediate situation, by providing food and preventing starvation, seems humane, the larger issue is in reality the quality of man's future genetic material, on which the survival of culture depends. Society should not through inaction adopt a course of negative eugenics, while rejecting a positive approach because of its controversial nature.

Some will argue that intelligence is not worth salvaging, that the more capable have produced destructive devices like atomic explosives, which threaten man's existence, and that the less able are likely to live in better harmony with nature. On this basis man should abandon the earth to lower creatures, like insects, but everyone is not that pessimistic. Man has evolved to a higher intellectual level, which is his only justification for a claim to superiority over other species and a consequent right to be the master of the world. There is no basic objection to continuing on that course, which nature herself selected in the first place. An artificial return to a lower grade of existence can be viewed as gross interference with natural processes, and such a course also would demand a great reduction in the size of human societies.

In reality the argument that scientists are menacing the world doesn't hold up to scrutiny. The actual scientific advances are potentially of great value, but they have been abused by politicians. The men who developed the theory behind the atom bomb did not intend for it to be used to threaten others, but the power to control the product was quickly usurped by political leaders, who perhaps put it to unwise use. Scientists need to assert their influence more positively and cease delivering their inventions for misuse by politicians with poor judgment. The problem of transferring increased power to men of wisdom, thus needs to be dealt with, although this is an issue which many populations hesitate to face.

Possibilities of genetic improvement

Since human heredity is no longer adequately guided by the natural forces that once charted man's future, it is now necessary to evaluate the newly created system and arrive at appropriate decisions. Even if the choice is to let things alone, allowing them to drift without active interference, this course is still an artificial one, in view of the changes that already have been introduced through social programs. In the past, women in the reproductive pool gave birth to numerous children, some societies averaging eight or more per family, one-third to one-half of those born failing to survive into adulthood. Although many deaths involved perfectly normal and perhaps superior children, a selection factor was obviously operating. In terms of intelligence this could be somewhat indirect, less able parents being more likely to lose their children to malnutrition and disease. Differential reproduction rates were also important, as established previously.

The present system of almost universal opportunity for reproduction, coupled with contraceptive limitation of family size and essentially total survival of children, has neutralized the former advantage of the more gifted, and there are indications of excessive reproduction now by the less able. The new system is thus far from being a natural one, and its merits must be evaluated.

Even the word "eugenics" is today frowned upon in social circles, although eminent geneticists point out that the eugenics approach cannot be viewed as a luxury, but rather should be seen as a course that is mandatory to insure continued adequacy of man's heredity. It would actually seem reasonable, in view of the changes that have been introduced, to at least attempt to take the necessary steps to make amends, so that those now living pass to the next generation material of equal quality to what they received. The soundness of that philosophy is obvious, while negative eugenics through drift makes no sense.

Unfortunately, there are serious conceptual problems with such a philosophy, even if it is accepted that society has a moral obligation to attempt somehow to balance the damage now being done. Scientists cannot agree among themselves what the most favorable course should be. In part this results from confusion about the nature of polygenic inheritance. There appears to be a consensus that intelligence is multigenic, and this leads to the conclusion that bright individuals should be encouraged to reproduce more than others. In this context most people seem to envisage a collection of favorable genes, which, with the application of the

Continued Integrity of Man's Genetic Heritage

proper system, would gradually concentrate in the population, leading eventually to an improved human constitution.

The early eugenists indeed tended to see the gene pool as grossly contaminated by abnormal genes that needed to be weeded out. This view led to a concept of purification, in some ways comparable to the establishment of pure stocks of domestic animals. Within this framework it seemed rather clear what needed to be done to ensure a better quality of life and establish a superior breed of man.

With the new formulation that hybrid vigor may be a vital force in human advancement, the basic principles become quite different. Here genetic purity is seen as undesirable, and continued progress is dependent on perpetuation of a polymorphic state. While the older concepts still are applicable to the elimination of rare inherited diseases, the larger problem of continued intellectual superiority must be approached from another point of view.

The approach of this book can help to clarify some of the fundamental problems. If it is correct that many, if not most, of the component factors of the proposed multigenic system of intelligence are in fact dependent on a carrier state, the ablest individuals are probably heterozygous with respect to several important genes. This in turn means that such persons cannot breed true. Serious disorders accompanying the superior state must be accepted as a part of the total package. This may be compared to the situation in Africa, where there was no way to produce a superrace resistant to malaria, the price of sickle-cell anemia having to be continuously paid by each generation. Similarly, it appears that abundant creative intelligence can be provided only by acceptance of a corresponding increase in mental illness, alcoholism, and myopia. The basic system does not lend itself to purification. Furthermore, only some of the heterozygous individuals exhibit the favorable trait in the case of intelligence. Hence the proper questions seem to be: Can society tolerate more mental illness in order to have increased creativity? Should man sacrifice his visual acuity to raise the overall IQ level? Should alcoholics be supported and their reproduction be encouraged in order to have more effective leaders? Without grappling with these issues there is no answer to the intelligence question. If the opportunity for a superrace actually had existed in the past, it probably would already have materialized somewhere, the better fit taking over the world, but the possibility of a perfect man is apparently only a myth.

Regrettably there is thus an unavoidable price for superiority. Society must accept defective individuals as a payment for superior minds, and gifted or creative persons must suffer their own illnesses and see their relatives often poorly equipped for life. The system has automatic built-in weaknesses, which cannot be eliminated. However, despite the fact that persons of superior ability often produce sick offspring, society appears to benefit from preservation of their genes in the population pool.

The example above, of the sickle-cell gene, may parallel more closely the situation in regard to the myopia gene in Eskimo populations than at first glance is apparent. In the sickle-cell case the existence of the gene enabled man to survive in otherwise uninhabitable regions by taking advantage of the superiority of the heterozygote. The most plausible explanation for the apparent high carrier rate for myopia among Eskimos is that here, too, there indeed is an advantage associated with the carrier state, most likely an intellectual one. At least in some Eskimo cultures young men are said to have been sent, when they reached maturity, to live alone for a year and prove their abilities to survive. Under harsh conditions this was probably tantamount to sending the myopes off to die, and perhaps the noncarriers could not survive either. Food supply has been so limited for these people that a group deciding to support its invalids could not compete. It seems cruel to abandon or even kill your aging parents, but at least some tribes of Eskimos are said to have considered failure to perform such acts a shirking of filial duty. Survival to age seventy years, when all offspring is likely to be grown, may be biologically adaptive for most human populations, but under the threat of starvation, Eskimos could not afford to abide by that principle. It is in areas of marginal livelihood that the operation of balanced polymorphisms becomes critical and most likely to be apparent.

Since it now appears that the disadvantages of myopia have been largely overcome, and in regard to this gene the additional benefit turns out to be that the homozygote actually enjoys the greatest brain enhancement, the possibility now may exist for an improved totally myopic human race. The first nation to promote such a development may outperform all others. Wise men need to contemplate whether active steps along such lines should be recommended. This is the only area where somewhat of a superrace now seems a possibility, and it results from the introduction of a medical device.

The low rate of mental deficiency among myopes also brings up an important issue. If a population existed somewhere where myopia was approaching a 100 percent rate, few institutions for the retarded would be required. This is a thought-provoking situation, which proponents of eugenics probably find intriguing. Perhaps the reports are correct that some Jewish and Chinese groups exhibit almost a 50 percent rate of myopia, so that more than one-half of the eye genes in these ethnic groups may already be of the myopia variety. The Chinese do claim to have low rates of mental deficiency.

There is a decided disadvantage to having to wear eyeglasses, especially among women. Even the adjective "myopic" has been assigned a second meaning, spelling out an implied criticism. But myopes can now pride themselves on being children of the future.

Social obligations of geneticists

Since apathy on the part of social leaders may be inviting genetic deterioration, there is a pressing need for geneticists to assert themselves and demand the influence that rightfully belongs in their domain. The findings described in this book appear to establish that the development of human intelligence is to a large extent dependent on genetic factors, which consequently must be taken into account in social planning. At present, there exists no human genetics department at any American university which is addressing itself to these fundamental issues, and geneticists are derelict in their duties not to demand action in this area. Leaders in this field should not be satisfied to just collect their salaries and then sit back and ignore the important concerns, which are vital to mankind's future. They need to be challenged to go into action and expose the fallacies which antigeneticists and politicians have foisted on the people. Unfortunately the tenure system for professors encourages them to stay away from controversial issues, but mankind can no longer afford that course.

Every large university should have a human genetics department, staffed by scientists who are prepared to make a bold attack on the problem of human intelligence. They should replace those who want to quietly play with flies and call it human genetics. Geneticists should also accept the responsibility of demonstrating to society how man's future genetic material must be properly safeguarded. New knowledge needs to be gathered to guide the wisest course, since mankind cannot just leave to arbitrary forces the fate of its genetic legacy. The survival of human culture will depend mainly on the brain quality of future men.

It also should be apparent that eugenists need to focus their efforts on studies of breeding patterns, rather than concentrating on a few individuals with grossly defective genes, which are being weeded out by natural selection anyway. The present interference with selective forces is what urgently needs attention.

Perhaps some of the university posts held today by psychologists should be transferred to geneticists, since the former have been proven wrong in their social interpretations. Psychology departments could at least hire geneticists to join their staffs. It seems absurd to require students to learn outmoded psychologic theories for college credit, while the important biologic principles are not even mentioned. A change in emphasis is obviously in order, for the good of all.

One must hope that the facts elaborated in previous chapters will not just be filed away to collect dust on library shelves. It may be correct that the truth eventually wins out, but that process can take thousands of years, and by then the opportunity may have been lost. No one is advocating genetic uniformity or standardization, but it is obviously unwise to leave the future to an artificial system of negative eugenics, just because geneticists are paralyzed by fear of criticism.

Suggestions have been made that society should be involved in plans for artificial insemination, at least in cases where this procedure is already being practiced. More knowledge is needed, however, before fully rational decisions are possible. Besides, the choices of the individual may not always coincide with the needs of society.

More distant issues involve possibilities which have not been fully developed, but which probably will come into existence before long. Thus the use of amniocentesis and other related developments will increasingly enable scientists to predict the quality of the unborn fetus, perhaps making possible the avoidance of the birth of those who are likely to be a burden on society. By promoting reproduction by the better fit and then preventing the birth of inferior offspring, it may be possible to insure better quality of the future genetic material. Mending after birth such defects as arise is far more difficult than providing for good material in the first place.

The principle of cloning is also being discussed. It is entirely conceivable that it may before long be possible to harvest ova from the female reproductive tract and after fertilization replace the nucleus with one from a cell of a superior person. Reimplantation of the egg than permits it to proceed with its development. This kind of procedure has been carried out successfully in lower animals, so that society may soon have to decide whether it wants a thousand Einsteins.

While these issues are rather frightening to contemplate, and particularly the matter of who decides who is worth preserving, the fact that they are threatening does not justify ignoring them. The questions are here, and they must be dealt with. Although inaction is politically expedient, it is not necessarily the best course, particularly since the current system is itself artificial.

It is not the intent of this discussion to make any specific recommendations other than increased exploration of such matters. But wise men should think about these issues, and face the facts. Genetic factors will determine the quality of man's future, and the system has already been disturbed. Perhaps it is possible to use reason, and positive steps are preferable. These matters should not be left to relatively uneducated politicians, who are mainly concerned with their own present material abundance, and society must avoid snap judgments which are not based on fact. Even then, mistakes will occur, but imperfect planning is certain to be better then none.

22
Future of Creative Intelligence

The genetic aspects of intellectual development have been ignored for decades, so that opportunities now abound for significant progress. Although unproven psychologic theories have dominated this field in recent years, a new day is at last dawning, demanding greater emphasis on biologic interpretations.

Recognition of scientific findings

Elucidation of the biologic substrate of creative intelligence remains an academic exercise as long as no use is made of the new information. A bold defense of heredity is imperative to overcome the very strong negative attitudes. Environmentalists managed to gain such a solid foothold because of the failure of scientists to stand up for their beliefs and express their convictions. At present there are few fields besides personality genetics where scientific evidence is absolutely overwhelming, yet its recognition almost totally blocked. Scholars must take the lead by insisting on a change in this atmosphere.

The opponents of the genetic view are, however, unlikely to relinquish without considerable resistance their control over the communal resources. The scientific basis for personality genetics has been measurably strengthened, although continued vigilance is needed to buttress the foundation further and ensure eventual recognition of the hereditary contribution. Nevertheless, in view of the predictable tendency for a rejection of significant changes, it is necessary, in addition to the scientific attack, to bring the subject to the attention of the public and make the proper appeals to political leaders.

Proponents of the environmental view have generally presented themselves in a humanitarian role, often labeling hereditarians as somehow evil or at best only pessimistic. In reality, however, this projection represents a reversal of the true situation, because artificial beneficence is only a deception, which helps no one in the long run. Only a realistic appraisal, based on a valid assessment, can have long-term value for mankind. Suppression of the facts does not alter them, and the time of reckoning may be harder to face if the truth has been compromised. Furthermore, since a delay in instituting necessary steps to deal with the actual situation is certain to lead to further deterioration, it is even possible that an irreversible course may be taken.

It takes an honest approach to man's problems to arrive at practical solutions. Denial of reality can lead to disaster, and proponents of such a denial are guilty of misleading the people, not those who deal in verifiable facts. Unfortunately, however, the masses tend to support leaders who tell them what they like to hear, even when it is a deliberate distortion.

Continued negativism toward heredity

The data from a voluminous study of the family distribution of IQ scores, conducted several years ago in The Netherlands, have been the subject of much discussion recently. The initial purpose of the investigation was to explore whether a well-documented period of famine, which occurred during the latter part of World War II, had adversely influenced the IQ of children exposed to intrauterine malnutrition. While a reduction in birth weights was demonstrable, no lowering of the adult IQ could be detected.

Incidental findings indicated that a very definitive inverse relationship existed between family size and IQ. Children, as well as parents, from large families had lower scores than those from smaller families, the progression being very orderly, with the only exception that single children rated lower than the eldest children

Future of Creative Intelligence

from two- to four-child families. Presumably only children included those who were illegitimate or derived from broken homes.

A second very important observation was that within each family size there was a systematic decline in IQ scores with successive pregnancies, again very orderly and, although not large in magnitude, definitively measurable with these huge samples of children. A still larger American study has led to essentially parallel data.

It is interesting to see how psychologists instinctively search for an environmental explanation for observations of this kind. Two separate groups have explored and widely circulated the hypothesis that the entire constellation of findings can be explained through a basic favorable parental environmental effect, which is progressively diluted as more children are added to the family. Furthermore, to explain the lower IQ of the parents in larger families, it is postulated that children in turn have an adverse effect on the parents' performance. Thus a multi-child family is visualized as a child-ish family, all members suffering from such an environment in their adult IQ measures.

Although one-child families do not fit into the total pattern, it is actually possible to explain the rest of the data with this environmental model. The first child in a family is presumed to benefit from intensive contact with the parents, experiencing negative influences from the next child, who in turn suffers even more. Each additional child brings the entire group down another step, landing himself on the bottom of the heap.

The principal weakness of this proposal is that there is no independent evidence that children influence each other adversely. Some would argue that a larger group should be of benefit through an enrichment of experiences. The hypothesis also fails to even consider the well-documented genetic factors, and obviously no explanation is adequate if it simply ignores a significant aspect of the overall problem, which has been scientifically proven to be important.

No unitary biologic interpretation has been proposed to explain the entire IQ distribution in these studies, but the findings would have been predicted on the basis of established scientific knowledge. To a geneticist it comes as no surprise that in Western societies there is now a trend for more intelligent parents to limit the number of their offspring, leading to smaller families attaining higher IQ scores. This is precisely what hereditarians have been trying to alert the public to, pointing out the lurking danger for the future of such societies.

Genetic explanations cannot account for the progressive decline in IQ in successive children in a family. However, biologic observations are available, which lead to a prediction of this phenomenon. Blood group incompatibilities, involving both the Rh factor and the ABO system, are known to cause progressively greater damage to the fetus as the number of pregnancies increases. Presumably there are additional factors of this type, which still have not been discovered. A biologist can thus predict an orderly progression of greater harm to the child in successive pregnancies, which would show up in studies of very large groups.

Since the biologic explanations are based on established phenomena, known to be operative and therefore no doubt responsible at least for a part of the observed differences, there is in reality no need for a contrived and entirely speculative environmental mechanism. Those who propose such explanations should be expected to furnish independent data in support of their theses. If their interpretation is correct, it should be devastating to a child to be raised in an orphanage or to be exposed to nursery schools or child-care centers. It would even be dangerous to adopt a second child as a companion for an only child. Isolation of children from each other has never been seen as a positive step, and it seems irresponsible to promote unsupported hypotheses which lead to that recommendation.

The refusal of the academic establishment to accept the evidence for a superior inborn mentality of myopes illustrates further the attitude which scientists have to contend with when exploring differences in human development. Investigators are not always extremely conservative in reaching conclusions, but in this field every other explanation has had to be given priority. When the scholarly community is totally disinclined to accept the most plausible explanation, it is always possible to claim that alternate possibilities have not been adequately ruled out.

In this atmosphere it is almost foolhardy to venture toward the next step, that of recommending that myopes should in some manner be encouraged to propagate. Less drastic proposals have inflamed the masses toward demonstrations and protests, demanding from political leaders a compromise or even a sacrifice of the welfare of future generations. But someone should stand up for unborn children, even though politicians find it expedient to ignore their rights. Their lives depend on the genes which children being born at present will hand down to them.

The scientific evidence is actually sufficiently strong to support a recommendation for action. If further evidence is desired, let society make funds available to gather the data, rather than just proclaiming that the knowledge is inadequate. More definitive verification is easily attainable if the desire is there to establish the scientific truth.

Since there is evidence of decreased reproduction now by the more gifted, presumably including myopes, it is likely that reduced incidence of myopia may be seen in the next generation. The facts in this matter could easily be ascertained, for example by making a survey of families now having children. But this is the type of information that political leaders prefer to sidestep.

Importance of genetic equilibria

The normal forces of natural selection are presumed to have promoted over the ages gradual improvement in man's genetic constitution. Although it is less recognized, these factors are of still greater importance as protectors of the integrity of the existing genetic endowment. In one aspect of that problem, mutation pressure constantly results in a deterioration of genes, and continued selection is necessary to counterbalance that force. Exposure to artificial radiation may in recent years have increased the load of undesirable mutations.

If it can be definitively established that hybrid vigor has indeed been an important catalyst in the attainment of higher human intelligence, the need for continued selection takes on still greater dimensions. To maintain a genetic polymorphism it is mandatory to ensure constant operation of selective forces. For example, the schizophrenia gene is always being weeded out, in part because the abnormal homozygotes become sick early in life and are unlikely to reproduce and also because even intermittently sick patients tend to have few children. In the past this loss has presumably been compensated for by the reduced ability of the "normal" homozygotes to compete in the struggle for survival, allowing the heterozygotes to persist at a relatively high frequency.

A preliminary study of the Icelandic population supports the notion that families with a schizophrenia gene indeed have in the past enjoyed an overall reproductive advantage. From a cohort of persons born in the interval 1881-1910 those sibships were selected which consisted of at least eight surviving children. Focusing on those among the offspring who fit into the period of study, it was possible to demonstrate that their psychosis rate was double that of the general population. Families with a schizophrenia gene thus appear to exhibit high fertility as well as an ability to provide for their children.

If the impression is correct that in present-day societies "normal" homozygotes have been not only released from the negative selection against them, but even encouraged to assume a leading role in the reproductive activity, it appears that gains made over many centuries may quickly be wiped out. Advocates of reduced population growth have welcomed recent signs that excessive multiplication may now be checking itself, but in reality it seems unwise to solve the overpopulation problem through elimination of the more capable segments of society. If this development is allowed to proceed unchecked, mankind may already be on a full-fledged course toward genetic deterioration.

The postulated polymorphisms thus carry extremely important implications for the future of man's heredity. The proper balance of genes in such systems can only be maintained by constant and relentless selection, relaxation of that pressure leading quickly to a loss of valuable mutant genes. Perhaps this mechanism is in part responsible for the recognized phenomenon that a human civilization declines when it has reached a state of material overabundance. When food supply no longer depends on competition, so that all fertile individuals are able to reproduce unimpeded by economic concerns, the quality of the genetic material is bound to deteriorate, if the composition of the biologic system has normally been regulated through survival of the more fit. This loss is further accelerated if the less intelligent are allowed to assume an artificially elevated reproductive rate.

Further evidence in support of a constant operation of a selective system has been encountered in independent studies of several populations. It has been known for some time that certain sectors of society have in the past enjoyed a survival advantage in a genetic sense, although no adequate explanation of the findings has hitherto been available. One study, done in Denmark, indicated that during the nineteenth century one-fourth of the original population remained highly represented, the other three-fourths essentially disappearing in the course of time. In three successive generations the successful original 25 percent rose to 50, 73, and 97 percent in terms of the fraction of new descendants derived from them. A similar study in France led to comparable results, showing only 12 percent of the population alive at the time of the great revolution being represented by living offspring a century later. These data indicate that one segment of the ancestral generations constantly gains ascendance.

One question raised in this connection is whether perhaps selective pressures created by the industrial revolution were exerting their influence during the period studied, negating the need for a biologic explanation. A shift in the population makeup, caused by artificial forces, could then conceivably be sufficient to account for the observed data. However, it turns out that corresponding changes were occurring in Iceland, where little industry then existed, people mainly living on farming, supplemented by some fishing. Among males born in the interval 1851-1880, 28 percent produced over 67 percent of the next generation, as is shown in Table 22-1. Since it can also be demonstrated that the socially most successful individuals were the ones who raised the largest number of children, fertility in the sense of capacity to reproduce was probably not the chief factor, but rather ability on the part of the head of the household to provide a livelihood and the parents' resourcefulness or creativity in dealing with problems.

Table 22-1

Displacement of the reproductively least successful one-third of males born in Iceland in 1851-1880 by the most successful one-third, in terms of the fraction of all surviving children produced by each group

Number of offspring	Original adult males		Children in new generation	
	Number	%	Number	%
0	443	32.7	0	0.0
1-4	533	39.3	1320	32.8
5-16	379	28.0	2700	67.2

The way one segment of society is actually expected to gradually replace the entire population, on the basis of a balanced polymorphic system, may be illustrated by the African example of sickle-cell anemia. Here the surviving individuals are either sickle-cell carriers or normals, as the abnormal homozygotes die without reproducing. The only source of carriers is other carriers. However, the latter replace not only themselves, but also a portion of the normal group. Assuming that the sickle-cell carriers remain a constant fraction, since the system has been balanced, the proportion of the total population derived from the carriers existing in an original generation chosen for study consequently increases with each passing period, until the initial carriers have achieved a total takeover. Although the situation may be somewhat more complex when one considers an intelligence gene or a system of several such genes, the overall expectation is the same, that one segment of the original people will be found to continually replace the total population.

If constant selective pressures indeed have operated during the entire course of human existence to maintain balanced polymorphic states, which are essential for continued superior intelligence, it becomes obvious that man cannot on his own initiative substitute an artificial system, which fails to take account of the basic requirements. Cultures which choose a course that disturbs the natural equilibria are bound to engineer their own decline. These factors should be seen as paramount by those concerned about man's future. It will do little good to make plans for economic abundance, if in the process the quality of human life itself is sacrificed.

Although specific data are at present available only with respect to the schizophrenia gene, to demonstrate how its carriers have been reproductively successful, it may be presumed that other genes involved in creative intelligence have been influenced by the same mechanisms. The alcohol gene has apparently been maintained at a high frequency by heterozygote advantage, but information to specifically document this has not been assembled. In the past it seems likely that myopia carriers also enjoyed a survival advantage, but now the homozygotes have taken over that function, having been released from the previous selection against them. Other frequent genes remain to be studied, including that presumed to be involved in diabetes.

The proposed relationship of superior intelligence to hybrid vigor, based on a heterozygous state at several genetic loci, requires a constant replacement of the total population by a sector which happens to carry the necessary genes, and there is no way that humans can thwart their biologic destiny. Competition seems to have been built into man's biologic structure, being therefore necessary to maintain proper balances. If it is correct, as recent statistics indicate, that Western societies are now characterized by overreproduction of the less intelligent, it is not hard to understand why the national scores on scholastic aptitude tests are showing a progressive decline. There is no need to postulate that schools have become inferior. Man cannot arbitrarily revamp his breeding structure without paying the associated price. These issues are not merely of academic interest, rather they deal with a grave potential peril that can threaten all humanity.

Sociologists should be trained in biologic principles, so that they become aware of the normal forces of nature and are equipped to provide rational leadership. Perhaps modern man can then, through use of his brain, avoid the type of decline that has previously characterized all advanced cultures. The naive view that mankind only needs to throw off its evil oppressors to attain a state of perfection, allowing equal reproduction by all, is so grossly misleading that those who actively promote such a course are themselves the real culprits. By distortions of the evidence, well meaning humanists can lead man toward oblivion.

Although in times past the question of actual survival of less capable individuals, in addition to differential reproductive rates, has been involved in maintenance of essential equilibria, it should be pointed out that destruction of the unfit is not necessary to protect the hereditary machinery. Genetic death means failure to reproduce, and it is entirely possible to keep the proper balances through adjustment of reproductive patterns. However, the present system of supplying everyone with a livelihood, without somehow regulating childbearing practices, leads to a violation of vital principles. Some system is needed which preserves the quality of the genes, while dealing humanely with those who should not produce offspring.

It must also be understood that failure to take action in this area is sure to lead to genetic decline. The concern is not about the distant future, but rather about changes that occur in a few generations. The old mechanisms have already been abandoned, and unless something is done soon, intellectual decline seems certain, leading to a demise of human culture. This kind of thinking will inevitably be labeled as unwarranted pessimism, and it doubtless will be rejected by the public, but wise men will see the validity of the arguments and the necessity to heed the warnings. There is nothing sinister about pointing out the dangers, in particular when a positive approach is available, which will effectively avert the potential disaster. Man needs to be protected not only from atom bombs and pollution, but also from his own follies in dealing with reproductive practices.

Mankind's long-term destiny

Because of the human tendency to resist changes, the only realistic expectation is that progress in human genetics is doomed to remain slow. It nevertheless seems important to at least try to direct the flow of events toward the area of greatest promise and ensure some gains, even though they must take their time.

While man's knowledge of the underlying processes gradually increases, biologic evolution will proceed on its course. New gene mutations are, however, so time consuming that they are probably of no consequence for man's foreseeable future, but alterations in the balances of presently existing differences can occur more rapidly.

Medical developments have generally been seen as eugenically negative. However, the successful treatment of myopia is an example of a therapeutic invention having assisted nature in overcoming a barrier which for some time had blocked the evolutionary thrust. Viewed as a brain gene, the myopia factor always had useful potentials, but it was kept in check by the associated eye defects, until these could be overcome. Perhaps there are other genes whose beneficial potentials can be similarly enhanced, enabling man to influence his own biologic improvement. It is not inconceivable that the schizophrenia homozygote may some day have the most effective brain. But in the main man is limited to making the best use of what already exists and guarding against unnecessary deterioration.

Man's future is largely dependent on the degree of wisdom that can be applied in the realm of reproduction. There has been much talk about the rights of the individual, but the basic idea of democracy, although obviously a noble one, may not be serving mankind favorably in the area of future genetic planning. One can legitimately question whether an unproductive person should have the same say on this matter as a highly creative scientist. There is obviously no logic to a democratic procedure which allows present populations to meddle unknowingly or unwisely with the genetic material of future generations.

The average person seeks to avoid boredom above all, and where the power is left in his hands, resources are expended on sports, comedies, and other forms of entertainment. Given the opportunity to regulate his own mind, even an intelligent person often seeks a state of increased arousal, which tends to be associated with a sensation of pleasure. A hypomanic level is likely to be what most persons would choose, if they could titrate their own mental state. This inclination obviously does not deal with future problems or take the next generation into account. The average person needs his arousal apparatus stimulated, and somehow this must be dealt with. But it seems wrong to exhaust the world's supply of valuable materials to entertain a group that came into existence by accident. Perhaps it would be preferable to ensure a better quality of genetic material in the first place, populating the world with people who do not need artificial entertainment.

The hypomanic state, while attractive to some and often effective as concerns personal enjoyment, socialization, and leadership inclination, does not seem to be what the world needs to enhance culture or creativity. In fact, the hypomanic person tends to be scatter-brained, unlikely to persevere in grappling with difficult problems. Individuals with a depressive mood often seem more effective in that regard. Many of the greatest scholars have suffered from a depressive tendency, but society has benefited from the tenacity which has frequently accompanied their somber temperament. While the hypomanic exhibits a higher energy level, flitting from one activity to another, the depressed person persists in his chosen endeavor, even accepting undesirable interruptions. Many productive scholars have considered themselves chronically ill, even spending much of their time in bed, but the creative fury somehow survived.

The various studies discussed earlier in this book converge to indicate that excessive arousal, perhaps through dopamine stimulation originating in the primitive areas of the brain, is intimately involved in alertness, wakefulness, increased brain activity, and creativity. These effects then interact with other stimulant influences. Perhaps in the future it will become possible to artificially regulate these systems so as to usefully enhance the function of the brain. However, it has been man's experience that each step in such endeavors is fraught with serious attendant risks. When an attempt is made to tinker with mechanisms which are closely connected to pleasure, sexual drive, or feelings of omnipotence, great care is necessary to avoid disaster.

The concept of genetic engineering promises to introduce a new dimension into human biologic evolution. If appropriate planning becomes possible, very significant shifts may in the future be made overnight by the use of such techniques. Only time will tell whether mankind has the necessary wisdom to make good use of such potentials. But guided evolution may some day replace spontaneous evolution, if proper judgment can be exercised. Further understanding of these factors may lead to increased control, but one must hope that man's free choice will not become his undoing. Creativity is a noble goal, but adventure must be tempered with caution when the stakes are so high. If good judgment prevails, and genetic decline can be avoided, man can continue to enjoy his increasing mastery of his world—the gift of his superior brain.

References

Chapter 1—Genetic diversity in human populations

Campbell, B. G. 1974. *Human Evolution.* 2nd ed. Aldine, Chicago.

Dobzhansky, T. 1970. *Genetics of the Evolutionary Process.* Columbia University Press, New York.

Chapter 2—Pedigree studies of common genetic systems

McKusick, V.A. 1969. *Human Genetics.* 2nd ed. Prentice-Hall, Englewood Cliffs, N.J.

Stern, C. 1973. *Principles of Human Genetics.* 3rd ed. Freeman, San Francisco.

Chapter 3—Quantitative family studies

Carter, C. O. 1969. The genetics of common disorders. *British Medical Bulletin* 25:52-57.

Gates, R. R. 1946, *Human Genetics.* Macmillan, New York.

Roberts, J. A. F. 1964. Multifactorial inheritance and human disease. *Progress in Medical Genetics* 3:178-216.

Whittinghill, M. 1965. *Human Genetics and Its Foundations.* Reinhold, New York.

Chapter 4—Estimates of heritability

Falconer, D. S. 1960. *Introduction to Quantitative Genetics.* Ronald Press, New York.

References

Feldman, M. W. and Lewontin, R. C. 1975. The heritability hang-up. *Science* 190:1163-1168.

Lerner, I. M. 1968. *Heredity, Evolution, and Society.* Freeman, San Francisco.

Chapter 5—Identification of useful mutant genes

Allison, A. C. 1964. Polymorphism and natural selection in human populations. *Cold Spring Harbor Symposia on Quantitative Biology* 29:137-150.

Huxley, J., Mayr, E., Osmond, H. and Hoffer, A. 1964. Schizophrenia as a genetic morphism. *Nature* 204:220-221.

Lee, C. C. 1961. *Human Genetics.* McGraw-Hill, New York.

Chapter 6—Biochemistry of gene function

Harris, H. (Ed.) 1975. *Principles of Human Biochemical Genetics.* 2nd ed. North-Holland, Amsterdam.

Stanbury, J. B. Wyngaarden, J. B. and Fredrickson, D. S. (Eds.) 1972. *The Metabolic Basis of Inherited Disease.* 3rd ed. McGraw-Hill, New York.

Watson, J. D. 1976. *The Molecular Biology of the Gene.* 3rd ed. Benjamin, New York.

Chapter 7—Definitions of intelligence

Anastasi, A. 1976. *Psychological Testing.* 4th ed. Macmillan, New York.

Fincher, J. 1976. *Human Intelligence.* Putnam, New York.

Matarazzo, J. D. and Wechsler, D. 1972. *Measurement and Appraisal of Adult Intelligence.* Williams and Wilkins, Baltimore.

Chapter 8—Hereditary contributions to learning ability

Dobzhansky, T. 1973. *Genetic Diversity and Human Equality.* Basic Books, New York.

Erlenmeyer-Kimling, L. and Jarvik, L. F. 1963. Genetics and intelligence: a review. *Science* 142:1477-1479.

Higgins, J., Reed, S. and Reed, E. 1962. Intelligence and family size: a paradox resolved. *Eugenics Quarterly* 9:84-90.

Honzik, M. P. 1957. Developmental studies of parent-child resemblance in intelligence. *Child Development* 28:215-228.

Munsinger, H. 1975. The adopted child's IQ: a critical review. *Psychological Bulletin* 82:623-659.

References

Skodak, M. and Skeels, H. M. 1949. A final follow-up study of one hundred adopted children. *Journal of Genetic Psychology* 75:85-125.

Thomas, A., Chess, S., Birch, H. G., Hertzig, M.E. and Korn, S. 1963. *Behavioral Individuality in Early Childhood.* New York University Press, New York.

Chapter 9—Evidence for a myopia gene

Dunphy, E. B. 1970. The biology of myopia. *New England Journal of Medicine* 283:796-800.

Furusho, T. 1957. Studies on the genetic mechanism of shortsightedness. *Japanese Journal of Ophthalmology* 1:185-190.

Hirsch, M. J. 1952. The change in refraction between the ages of 5 and 14. *American Journal of Optometry* 29:445-459.

Karlsson, J. L. 1974. Concordance rates for myopia in twins. *Clinical Genetics* 6:142-146.

Karlsson J. L. 1975. Evidence for recessive inheritance of myopia. *Clinical Genetics* 7:197-202.

Morgan, R. W., Speakman, J. S. and Grimshaw, S. W. 1975. Inuit myopia: an environmentally induced epidemic? *Canadian Medical Association Journal* 112:575-577.

Young, F. A. 1963. The effect of restricted visual space on the refractive error of the young monkey eye. *Investigative Ophthalmology* 2:571-577.

Chapter 10—Myopia and intelligence

Grosvenor, T. 1970. Refractive state, intelligence test scores, and academic ability. *American Journal of Optometry* 47:355-361.

Hirsch, M. J. 1959. The relationship between refractive state of the eye and intelligence test scores. *American Journal of Optometry* 36:12-21.

Karlsson, J. L. 1973. Genetic relationship between giftedness and myopia. *Hereditas* 73:85-88.

Karlsson, J. L. 1975. Influence of the myopia gene on brain development. *Clinical Genetics* 8:314-318.

Chapter 11—Existence of an alcoholism gene

Amark, C. 1951. *A Study in Alcoholism.* Acta Psychiatrica Scandinavica, Supplement 70.

Bourne, P. G. and Fox, R. (Eds.) 1973. *Alcoholism. Progress in Research and Treatment.* Academic Press, New York.

Goodwin, D. W., Schulsinger, F. and Hermansen, L. 1973. Alcohol problems in adoptees raised apart from alcoholic biological parents. *Archives of General Psychiatry* 28:238-243.

Kaij, L. 1960. *Alcoholism in Twins: Studies on the Etiology and Sequels of Abuse of Alcohol.* Almquist and Wiksell, Stockholm.

Partanen, J., Bruun, K. and Markkanen, T. 1966. *Inheritance of Drinking Behavior.* Finnish Foundation for Alcohol Studies, Helsinki.

Roe, A. and Burks, B. 1945. *Adult Adjustment of Foster Children of Alcoholic and Psychotic Parentage and the Influence of the Foster Home.* Quarterly Journal of Studies on Alcohol, New Haven.

Schuckit, M. A., Goodwin, D. A. and Winokur, G. 1972. A study of alcoholism in half siblings. *American Journal of Psychiatry* 128:1132-1136.

Chapter 12—Brain stimulation associated with alcoholism

Cantwell, D. (Ed.) 1975. *The Hyperactive Child.* Spectrum Publications, New York.

Crowe, R. R. 1974. An adoption study of antisocial personality. *Archives of General Psychiatry* 31:785-791.

Goodwin, D. W., Schulsinger, F. and Winokur, G. 1975. Alcoholism and the hyperactive child syndrome. *Journal of Nervous and Mental Disease* 160:349-353.

Lange, J. 1930. Crime as Destiny. Boni, New York.

Robins, L. N. 1966. *Deviant Children Grown Up.* Williams and Wilkins, Baltimore.

Sinclair, U. B. 1956. *The Cup of Fury.* Channel Press, Great Neck.

Chapter 13—Physiology of learning functions

Bedrossian, R. H. 1971. The effect of atropine on myopia. *Annals of Ophthalmology* 3:891-897.

Pribram, K. H. and Broadbent, D. (Eds.) 1970. *Biology of Memory.* Academic Press, New York.

Wallace, P. 1975. Neurochemistry: unraveling the mechanism of memory. *Science* 190:1076-1078.

Warburton, D. M. 1975. Brain, Behavior and Drugs. *Introduction to the Neurochemistry of Behavior.* John Wiley and Sons, New York.

Chapter 14—Definition of creativity

Barron, F. 1963. *Creativity and Psychological Health.* Van Nostrand, Princeton.
Ghiselin, B. 1952. *The Creative Process.* University of California Press, Berkeley.
Goertzel, V. and Goertzel, M. G. 1962. *Cradles of Eminence.* Constable, London.
Koestler, A. 1964. *The Act of Creation.* Macmillan, New York.
Lange-Eichbaum, W. 1931. *The Problem of Genius.* Kegan Paul, London.
Lombroso, C. 1891. *The Man of Genius.* Walter Scott, London.
Terman, L. M. 1925-59. *Genetic Studies of Genius,* Volumes I-V. Stanford University Press, Stanford.
Wallas, G. 1926. *The Act of Thought.* Harcourt Brace, New York.

Chapter 15—Evidence for inheritance of creativity

Andreasen, N. J. C. and Canter, A. 1974. The creative writer: psychiatric symptoms and family history. *Comprehensive Psychiatry* 15:123-131.
Dudek, S. Z. 1970. The artist as a person. Generalizations based on Rorschach records of writers and painters. *Journal of Nervous and Mental Disease* 150:232-241.
Galton, F. 1869. *Hereditary Genius.* Macmillan, London.
Hirsch, N. D. M. 1931. *Genius and Creative Intelligence.* Philosophical Library, New York.
Kretschmer, E. 1931. *The Psychology of Men of Genius,* Kegan Paul, London.
MacKinnon, D. W. 1965. Personality and the realization of creative potential. *American Psychologist* 20:273-281.

Chapter 16—Existence of a schizophrenia gene

Bellak, L. and Loeb, L. 1969. *The Schizophrenic Syndrome.* Grune and Stratton, New York.
Cunningham, L., Cadoret, R. J., Loftus, R. and Edwards, J. E. 1975. Studies of adoptees from psychiatrically disturbed biological parents. *British Journal of Psychiatry* 126:534-549.
Fischer, M. 1971. Psychoses in the offspring of schizophrenic monozygotic twins and their normal cotwins. *British Journal of Psychiatry* 118:43-52.

References

Heston, L. L. 1966. Psychiatric disorders in foster home reared children of schizophrenic mothers. *British Journal of Psychiatry* 112:819-825.

Kallmann, F. J. 1938. *The Genetics of Schizophrenia.* Augustin, New York.

Karlsson, J. L. 1966. *The Biologic Basis of Schizophrenia.* Thomas, Springfield.

Karlsson, J. L. 1970. The rate of schizophrenia in foster-reared close relatives of schizophrenic index cases. *Biological Psychiatry* 2:285-290.

Rosenthal, D. (Ed.) 1963. *Genain Quadruplets.* Basic Books, New York.

Rosenthal, D. 1972. Three adoption studies of heredity in the schizophrenic disorders. *International Journal of Mental Health* 1:63-75.

Slater, E. and Cowie V. 1971. *The Genetics of Mental Disorders.* Oxford University Press, London.

Zerbin-Rüdin, E. 1972. Genetic research and the theory of schizophrenia. *International Journal of Mental Health.* 1:42-62.

Chapter 17—Genetic transmission of psychotic tendency

Böök, J. A. 1953. A genetic and neuropsychiatric investigation of a north-Swedish population. *Acta Genetica et Statistics Medica* 4:1-100.

Cammer, L. 1970. Schizophrenic children of manic-depressive parents. *Diseases of the Nervous System* 31:177-180.

Elston, R. C. and Campbell, M. A. 1970. Schizophrenia: evidence for the major gene hypothesis. *Behavior Genetics* 1:3-10.

Helgason, T. 1964. *Epidemiology of Mental Disorders in Iceland.* Acta Psychiatrica Scandinavica, Supplement 173.

Kallmann, F. J. 1953. *Heredity in Health and Mental Disorder.* Norton, New York.

Karlsson, J. L. 1973. An Icelandic family study of schizophrenia. *British Journal of Psychiatry* 123:549-554.

Karlsson, J. L. 1974. *Inheritance of Schizophrenia.* Acta Psychiatrica Scandinavica, Supplement 247.

Shields, J. 1968. Summary of the genetic evidence. *Journal of Psychiatric Research,* Supplement 1:95-126.

Slater, E. 1958. The monogenic theory of schizophrenia. *Acta Genetica et Statistica Medica* 8:50-56.

References

Chapter 18—Relation of creativity to the schizophrenia gene

Bell, E. T. 1937. *Men of Mathematics.* Simon and Schuster, New York.
Ellis, H. A. 1926. *A Study of British Genius.* Houghton Mifflin, New York.
Hasenfus, N. and Magaro, P. 1976. Creativity and schizophrenia: an equality of empirical constructs. *British Journal of Psychiatry* 129: 346-349.
Juda, A. 1949. The relationship between highest mental capacity and psychic abnormalities. *American Journal of Psychiatry* 106: 296-307.
Lange-Eichbaum, W. 1928. *Genie, Irrsinn und Ruhm.* Reinhardt, München.
Lucas, C. J. and Stringer, P. 1972. Interaction in university selection, mental health and academic performance. *British Journal of Psychiatry* 120: 189-195.
Magoun, H. W. 1969. *The Waking Brain.* Thomas, Springfield.
Nisbet, J. F. 1900. *The Insanity of Genius.* Grant Richards, London.
Oden, M. H. 1968. The fulfillment of promise: 40 year follow-up of the Terman gifted group. *Genetic Psychology Monographs* 77: 3-93.
Saletu, B., Saletu, M., Marasa, J., Mednick, S. and Schulsinger, F. 1975. Acoustic evoked potentials in offspring of schizophrenic mothers. *Clinical Electroencephalography* 6: 92-102.
Thomas, H. and Thomas, D. L. 1940-47. *Living Biographies.* Halcyon, Garden City.

Chapter 19—Physiologic regulation of brain arousal

Celesia, G. and Barr, A. N. 1970. Psychosis and other psychiatric manifestations of levodopa therapy. *Archives of Neurology* 23: 193-200.
Connell, P. H. 1958. *Amphetamine Psychosis.* Chapman and Hall, London.
Cooper, J. R., Bloom, F. E. and Roth, R. H. 1974. *The Biochemical Basis of Neuropharmacology.* 2nd ed. Oxford University Press, New York.
Goodman L. and Gilman A. (Eds.) 1975. *The Pharmacological Basis of Therapeutics.* 5th ed. Macmillan, New York.

References

May, P. R. A. 1968. *Treatment of Schizophrenia.* Science House, New York.

Rockwell, F. V. 1948. Dibenamine therapy in certain psychopathologic syndromes. *Psychosomatic Medicine* 10: 230-237.

Seeman, P. and Lee, T. 1975. Antipsychotic drugs: direct correlation between clinical potency and presynaptic action on dopamine neurons. *Science* 188: 1217-1219.

Snyder, S.H. 1974. *Madness and the Brain.* McGraw-Hill, New York.

Stefano, G. B., Catapane, E. and Aiello, E. 1976. Dopaminergic agents. Influence on serotonin in the molluscan nervous system. *Science* 194: 539-541.

Chapter 20—Inherited personality and creative aptitude

Anastasopoulos, G. and Photiades, H. 1962. Effect of LSD-25 on relatives of schizophrenic patients. *Journal of Mental Science* 108:95-98.

Dalen, P. 1975. *Season of Birth. A Study of Schizophrenia and Other Mental Disorders.* North-Holland Publishing Company, Amsterdam.

Dalton, K. 1976. Prenatal progesterone and educational attainments. *British Journal of Psychiatry* 129: 438-442.

Fowler, R. C. and Tsuang, M. T. 1975. Spouses of schizophrenics: a blind comparative study. *Comprehensive Psychiatry* 16:339-342.

Hare, E. H. 1976. The season of birth of siblings of psychiatric patients. *British Journal of Psychiatry* 129: 49-54.

Janowsky, D. S., El-Yousef, M. K., Davis, J. M. and Sekerke, J. 1973. Parasympathetic suppression of manic symptoms by physostigmine. *Archives of General Psychiatry* 28: 542-547.

Kretschmer, E. 1926. *Physique and Character.* Harcourt Brace and World, New York.

Malzberg, B. 1940. *Social and Biological Aspects of Mental Disease.* State Hospitals Press, Utica.

McConaghy, N. and Clancy, M. 1968. Familial relationships of allusive thinking in university students and their parents. *British Journal of Psychiatry* 114: 1079-1087.

Morrison, J. R. 1974. Bipolar affective disorder and alcoholism. *American Journal of Psychiatry* 131: 1130-1133.

Murawski, B. J., Chazan, B. I., Balodimos, M. C. and Ryan, J. R. 1970. Personality patterns in patients with diabetes mellitus of long duration. *Diabetes* 19: 259-263.

Murphy, D. L. and Weiss, R. 1972. Reduced monoamine oxidase activity in blood platelets from bipolar depressed patients. *American Journal of Psychiatry* 128: 1351-1357.

Schildkraut, J. J., Herzog, J. M., Orsulak, P. J., Edelman, S. F., Shein, H. M. and Frazier, S. H. 1976. Reduced platelet monoamine oxidase activity in a subgroup of schizophrenic patients. *American Journal of Psychiatry.* 133: 438-440.

Thaler, J. S. and Goldberg, S. A. 1972. A vision census and study of a U.S. veterans administration hospital for psychiatric patients. *American Journal of Optometry* 49:796-800.

Young, F. A., Singer, R. M. and Foster, D. 1975. The psychological differentiation of male myopes and nonmyopes. *American Journal of Optometry and Physiological Optics* 52: 679-686.

Chapter 21—Continued integrity of man's genetic heritage

Allison, A. C. 1954. Protection offered by sickle-cell trait against subtertian malarial infection. *British Medical Journal* 1: 290-294.

Bajema, C. J. 1966. Relation of fertility to educational attainment in a Kalamazoo public school population. *Eugenics Quarterly* 13: 306-315.

Osborn, F. 1968. *The Future of Human Heredity.* Weybright and Talley, New York.

Young, F. A., Leary, G. A., Baldwin, W. R., West, D. C., Box, R. A., Harris, E. and Johnson, C. 1969. The transmission of refractive errors within Eskimo families. *American Journal of Optometry* 46: 676-685.

Chapter 22—Future of creative intelligence

Belmont, L. and Marolla, F. A. 1973. Birth order, family size, and intelligence. *Science* 182: 1096-1101.

Breland, H. M. 1974. Birth order, family configuration, and verbal achievement. *Child Development* 45: 1011-1019.

Goldberg, E. M. and Morrison, S. L. 1963. Schizophrenia and social class. *British Journal of Psychiatry* 109: 785-802.

References

Goldschmidt, E. (Ed.) 1963. *The Genetics of Migrant and Isolate Populations*. Williams and Wilkins, Baltimore.

Zajonc, R. B. 1976. Family configuration and intelligence. *Science* 192: 227-236.

Index

Abstinence, 87
Acetylcholine, 95, 144, 145, 164
Achievement, 73, 89, 169
Acrocephalosyndactyly, 35
Adaptation, 2, 4, 5, 37, 51, 58, 134
  group, 4, 8, 38, 134
  individual, 5
Addiction, 82, 87, 96
Adoptee, 116
Adoption, 30, 116
Adrenal gland, 95
Adrenalin, 146, 148
Africa, 36, 64, 173
Aggressiveness, 36, 104
Agriculture, 3
Alaska, 63
Alcohol, 80, 88, 163
Alcoholism, 62, 79, 87, 95, 157
Allele, 22, 43, 44
Allusive thinking, 158
Amark, C., 82, 83, 90
Amino acid, 39, 42
Amniocentesis, 176
Amphetamine, 147, 149, 150
Amphetamine psychosis, 147, 149
Andreasen, N.J.C., 106
Anhedonia, 108
Anthropology, 3
Appetite, 26
Architects, 105
Arousal, 94, 131, 133, 143, 145, 146, 150, 189
Ascertainment, 114
Association, 98, 101, 110
Atropine, 95
Autonomic nervous system, 132, 145
Axial length, 69

Bach family, 104
Barron, F., 100
Berkeley, California, 74, 160
Bernoulli family, 104
Bioassay, 40

Biochemistry, 9, 39
Blood group, 27, 182
Bohr family, 104
Böök, J. A., 126, 127
Brain, 3, 5, 87, 143, 144, 190
Branch of family, 110, 122, 123
Burks, B., 80

California, 65, 74, 76, 77, 116
Cambridge, 105
Canada, 63
Cantwell, D., 88
Carrier, 13, 127, 128, 132
Catecholamine, 146, 148,
Celibacy, 100
Central nervous system, 77, 143
Central stimulant, 94, 146
Cerebral cortex, 132, 145
Characteristic, 2, 9, 13, 17, 25
  continuous, 18
  hereditary, 9
  metric, 17, 18
  quantitative, 18
Chromosome, 9, 12, 85
Churchill, W., 50
Clinician, 22, 23
Cloning, 177
Cocaine, 147
Cognitive ability, 51
Color blindness, 95
Communication, 3, 4, 5, 38, 143, 155
Computer, 15, 21, 93
Conceptual thinking, 51
Concordance, 28, 29, 56, 65, 77, 80, 113, 114
Correlation, 28, 31, 48, 56, 57, 59, 60, 61
Cotwin, 20, 28, 55
  dizygotic, 113
  monozygotic, 20, 55, 113
Creativity, 97, 98, 103, 131, 133, 134, 138, 139, 158, 173
Crime, 89, 152
Criminality, 89, 168

Index

Culture, 3, 8, 25, 51, 78, 90, 171, 188
Cunningham, L., 117

Darwin, C., 139, 141
DeForest, L., 135, 138
Delirium, 134
Delusion, 108
Dementia praecox, 107
Denmark, 81, 115, 185
Depression, 108, 109, 161
Development, 1, 4, 12, 73, 163
    abnormal, 34
    personality, 29, 132, 160
Diabetes, 156, 160, 168
Diagnosis, 110, 126
Dibenamine, 151
Diopter, 65, 66, 68, 69, 76, 77
Disorder, 11, 12, 13, 14
    common, 34, 168
    dominant, 11, 13, 35, 42, 43
    recessive, 13, 35, 43
Diversity, 1, 8, 155
Divorce, 89
DNA, 39, 43
Dominance, 11, 13, 14, 88
Dopamine, 145, 146, 147, 148, 190
Dudek, S. Z., 106
Dumas family, 104
Dwarfism, 17, 33, 35

Ecstacy, 98, 135
Ectoderm, 147
Einstein, A., 5, 16, 139, 165
Electroencephalography, 131, 159
Eliot, T. S., 135
Ellis, H. A., 135
Emmetrope, 74
Environment, 3, 32, 133
Environmentalist, 81, 117, 179
Enzyme, 10, 12, 37, 41, 42, 44
Epidemic, 63, 70
Epinephrine, 146, 148
Equilibrium, 35, 183, 186
Eskimo, 26, 63, 65, 70, 78, 174
Etiology, 12, 118
Eugenics, 167, 169, 171, 172
Eugenist, 173, 176

Euphoria, 108
Europe, 64, 109
Evolution, 4, 8, 145, 147, 155, 188
Executive, 79
Expressivity, 18
Extraversion, 58, 161
Eye color, 11, 17, 29
Eye strain, 63

Famine, 180
Faraday, M., 138, 139
Female, 80, 126, 177
Fetus, 29, 177, 182
Finland, 80
Firstborn, 104, 105, 116
Fitness, 8, 78
Flatworm, 94
Fossil, 3
France, 161, 185
Fremming, 82
Frost, R., 135
Furusho, T., 67

Galton, F., 58, 104, 105
Gene frequency, 10, 33, 35, 87, 122, 128
Genes, 3, 9, 10
    ancestral, 10
    codominant, 10, 14, 43, 44, 61, 68
    dominant, 10, 20
    functional, 11
    mutant, 4, 15, 35, 43
    native, 12, 22, 34
    non-functional, 4, 6, 10, 34
    recessive, 10, 11
    sex-linked, 9
Genetic engineering, 190
Geneticist, 1, 7, 16, 61, 82, 175
Genetics, 1, 2, 9, 11, 17, 22
    biochemical, 40
    Mendelian, 9
    population, 21
Genius, 97, 100, 102, 103, 104, 134
Genotype, 12, 13, 14, 22
    recessive, 12
Germany, 67, 68
Ghiselin, B., 99
Giftedness, 48, 49, 51, 91, 98, 105

Index

Goertzel, M. G., 99
Goertzel, V., 99
Graduate, honor, 74
Gravity, 159
Greece, 104
Greek, 134
Grosvenor, T., 74

Half-sib, 19, 81, 112
Hallucination, 108, 110
Hallucinogenic agent, 146, 159
Hardy-Weinberg law, 22, 128
Hebephrenia, 162
Helgason, T., 83, 126
Hemoglobin, 43
Heritability, 25, 31, 32, 57, 58, 59, 80, 88
Heritage, 3, 5, 167
Heston, L. L., 117
Heterozygote, 6, 11, 14, 61, 174, 183
Hirsch, M. J., 74, 76, 77
Homeostasis, 143, 145
Homozygote, 7, 11, 14, 35, 36, 42, 44, 128, 163, 174, 183, 184, 186
  abnormal, 7, 11, 14, 36, 42, 44
  recessive, 35
Honzik, M. P., 56, 57
Hormone, 95, 161, 164
Hubble, E., 136
Humanist, 187
Huntington's chorea, 35
Hybrid vigor, 6, 7, 36, 155, 173, 183, 186
Hydroanencephaly, 30
Hyperactivity, 87, 88, 89
Hypermetrope, 74
Hypertension, 95, 168

Iceland, 54, 55, 58, 59, 74, 78, 82, 83, 84, 85, 90, 91, 116, 117, 122, 124, 125, 127, 169, 185
Inborn error, 39, 44
Incompatibility, 182
Incubation, 98
Index case, 13, 14, 19, 20, 21, 22, 29, 30, 31, 55, 60, 65, 68, 80, 88, 90, 110, 111, 117
Indian, 63
Industry, 3

Inheritance, 16, 20, 39, 121
  dominant, 12, 20
  polygenic, 18, 20, 59, 84, 131, 172
  recessive, 12, 20, 21, 68, 69, 70, 71, 116, 126
  sex-linked, 12, 66
Insanity, 135
Inspiration, 98, 134
Institution, 108, 140
Intelligence, 47, 73, 78, 103
Intelligence quotient, 48, 49
Intelligence test, 74, 90, 97
Intrauterine factor, 29, 117
Intuition, 99
Iowa, 106
Irish, 157

Japan, 67
Juda, A., 135
Judgment, 50, 51, 164

Kallmann, J. L., 112, 114, 115, 117, 118, 126, 127
Kindred, 84, 85
Kinship, 26, 84, 104, 122
Koestler, A., 99
Kraepelin, 109
Kretschmer, E., 135

Lange, J., 89
Lange-Eichbaum, W., 99, 135, 136
Language, 25, 51
Laterality, 29
L-dopa, 147, 148
Leader, 89, 97, 157, 160
Leadership, 58, 79, 91, 159
Learning, 53
Lee, T., 149
Leprosy, 25
Lesion, 144
Lewis, S., 136
Limbic system, 132
Lincoln, A., 141
Lithium, 152
Lobotomy, 153
Locus, 9, 10, 35, 124, 127, 187
Lombroso, C., 100, 135

Index

London, 136
Longevity, 100, 168
Lorge-Thorndike test, 76
Lovibond test, 158
LSD, 159

MacKinnon, D. W., 100, 105
Madness, 105, 134
Malaria, 36, 173
Malnutrition, 172
Malzberg, B., 157
Manic depressive illness, 108, 139, 152
Masters, E. L., 136
Mathematics, 139
Mayer, R., 139
Mednick, 131
Melancholia, 113
Melanin, 147, 148
Mendel, G., 15, 139, 141
Mendelian laws, 12
Metabolite, 6, 37, 40
Minimal brain dysfunction, 87
Mississippi, 135
MMPI test, 106, 158
Mongolism, 76
Monkey, 64
Monoamine oxidase, 162
Monomania, 100
Munsinger, H., 56, 57
Mutation, 4, 10, 35, 40, 44, 188
   induced, 40
   recurrent, 43
   spontaneous, 10
Mutation pressure, 34, 36, 183
Myelinization, 29
Myope, 67, 68, 69, 73, 74, 76, 163
Myopia, 62, 63, 66, 73, 95, 134

Napa, California, 65, 116
Natural selection, 6, 7, 8, 34, 35, 36, 70, 183
Nearsightedness, 26, 62, 63, 73, 77
Netherlands, 180
Neuroanatomy, 144, 147
Neurochemistry, 144
Neurohormon, 41, 95, 144, 146, 148
Neurologist, 144

Neurosis, 107
New Zealand, 74
Newton, I., 138, 165
Nisbet, J. F., 135
Nobel prize, 91, 153
Noradrenalin, 145
Norepinephrine, 145
Nucleus, 9

Originality, 165
Orphanage, 55, 182
Otis test, 74
Overabundance, 184
Overarousal, 131
Overinclusion, 100
Overpopulation, 184
Overstimulation, 88

Parkinsonism, 147
Paul, 67
Pedigree, 9, 12, 15, 16, 20, 26, 53, 68, 69, 110, 111, 122
Penetrance, 13, 14, 18, 21, 67, 68, 71, 84, 88, 128
   incomplete, 13, 14, 18
Personality, 155, 156. See also Development, personality; Trait, personality
Pharmacotherapy, 140, 151
Phenothiazine, 148, 150
Phenotype, 13, 20
Phenylketonuria, 42, 43, 44
Physics, 139
Physiology, 93, 143, 146
Physostigmine, 164
Platelet, 162
Pleiotrophy, 61
Politician, 53, 79, 171, 183
Polymorphism, 34, 36, 37, 61, 66, 73, 124, 174, 183
Polypeptide, 9, 95
Porphyria, 42
Pound, E., 136
Proband, 20, 28, 55, 115, 125
Professional, 89, 162, 164
Prognosis, 109
Prospective study, 30

Index

Protein, 39, 43, 94
Pseudodominance, 13, 66
Psychiatrist, 81, 128
Psychologic test, 100, 101, 133
Psychologist, 47, 53, 71, 97, 106, 128, 133, 153, 157, 176
Psychosis, 107, 109, 110, 126, 131, 133, 134
Psychotherapy, 151, 152, 153
Puberty, 64

Quackery, 140
Quadruplet, 115

Raven matrices test, 74
Receptor, 41, 149, 150
Relative, 19, 55, 60, 83
   first-degree, 19, 20, 55, 83, 113, 127
   second-degree, 19, 20, 55, 83
   third-degree, 19, 83
Reticular activating system, 131
Reykjavik, Iceland, 54, 55, 82, 90, 125
Rh factor, 182
RNA, 39, 94
Rockwell, F. V., 151
Roe, A., 80
Rorschach test, 101, 106
Rosenthal, D., 117
Royce, J., 136

Sampling, 13, 68, 113
Schizophrenia, 107, 109, 131
Scholar, 103, 179
Scholastic aptitude test, 187
School myopia, 64
Scientist, 25, 99, 138, 171, 182
Seeman, P., 149
Shields, J., 110
Sib, 19, 26, 115, 117
Sibship, 65
Sickle-cell, 36, 43, 173, 184
Sinclair, U.B., 90
Skeels, H.M., 56, 57
Skin color, 18, 19
Skodak, M., 56, 57
Slater, E., 127
Sociologist, 187

Sociopath, 157
Specialization, 4
Standard deviation, 49
Statistical method, 18, 21
Stature, 18, 33, 161
Stepparent, 81
Sterility, 4
Student, honor, 133
Suicide, 108, 133
Superrace, 173, 175
Survival, 2, 5, 8, 22, 61, 70, 78, 171, 184
Sweden, 80, 82
Symptom, 108, 149

Tension, 84
Terman, L. M., 74, 133
Threshold effect, 19
Trait, 1, 12, 17, 19, 23, 26, 29
   dominant, 11, 12
   metric, 22, 28, 33
   personality, 1, 15, 19, 25, 26, 28, 33, 36, 98, 161
   physical, 25
Tranquilizer, 149, 150, 152
Transmission, 12, 13, 14, 20, 27, 59, 66
   dominant, 12, 20, 66, 126
   polygenic, 20
   recessive, 13, 66
Triplet, 115
Triplet code, 39
Tuberculosis, 26
Twain, M., 136
Twin, 19, 27, 31, 53, 56, 64, 66, 77, 80, 87, 89, 104, 113, 115, 119
   dizygotic, 19, 27, 28, 29, 113
   fraternal, 19, 27
   identical, 19, 27
   monozygotic, 19, 27, 28, 29, 113

Unit character, 11, 17
United States, 64, 79, 80, 81, 109, 135, 170

Van Gogh, V., 141
Variability, 10
Variation, 14, 18, 32
Verbal fluency, 90

Index

Vertebrate, 132, 143, 145
Vision, 63, 64, 74

Wallas, G., 98
Wisdom, 160, 165, 190

Withdrawal, 108
Writers, 105, 106

Zerbin-Rüdin, E., 112, 127
Zygosity, 27

About the Author

A native of Iceland, Dr. Jon L. Karlsson received his B.S. degree in genetics and a Ph.D. degree in biochemical genetics from the University of California at Berkeley. He earned an M.D. degree at the University of California School of Medicine in San Francisco and was certified as a children's physician by the American Board of Pediatrics.

Dr. Karlsson is now one of the world's leading authorities on the inheritance of personality traits, having pioneered in the development of new approaches to research in that area. His accomplishments include the identification of several specific genes which influence mental development. He is continuing his research as a Consultant at the Human Genetics Laboratory, National Archives, Reykjavik, Iceland.

WITHDRAWN

Karlsson, Jon L.

Inheritance of creative intelligence /

~~LIBRARY~~
~~FLAGLER COLLEGE~~